Hard Times and a Nickel a Bucket

Hard Times and a Nickel a Bucket
Struggle and Survival in North Carolina's Shrimp Industry

John R. Maiolo
Professor Emeritus
East Carolina University
Greenville, North Carolina

CHAPEL HILL
PRESS, INC.

Copyright © 2004 John R. Maiolo

All rights reserved. No part of this book may be used, reproduced
or transmitted in any form or by any means, electronic or mechanical,
including photograph, recording, or any information storage or
retrieval system, without the express written permission
of the author, except where permitted by law.

ISBN 1-880849-77-1
Library of Congress Catalog Number 2004100293

Printed in the United States of America
08 07 06 05 04   10 9 8 7 6 5 4 3 2

To my wife, Ruth

Heading Shrimp for a Nickel a Bucket.

Contents

Preface ix

Acknowledgements xiii

Introduction to the Fishery 1

From Leafy Weirs to Otter Trawls 14

Laying on a Set: The Art of Channel-netting 34

Shrimping Comes of Age in North Carolina 46

Management: From Local Customs to Statewide Regulations 64

The Commercial Man 83

The Carolina Cannonball 115

The Struggle for Survival in a Changing Coastal Culture 146

Endnotes 179

Bibliography and References 183

About the Author 191

Preface

Among the various types of seafood, there is none more popular among American consumers than that small, tasty, crustacean, the shrimp. Its popularity, however, had a much later commercial beginning than that of other marine species in North Carolina, which date back to the War Between the States. Just after the turn of the twentieth century, many commercial fishermen still considered shrimp to be *pests* that fouled their nets while they hunted for other species! Serious commercial interest in shrimping in the United States developed during the first quarter of that century, gradually becoming a well-embedded feature of maritime communities and an important economic activity nationally, as well.

When harvesting shrimp became a viable commercial activity in North Carolina in the early thirties, fishermen were paid three cents a pound. The women who *headed* the shrimp were paid a nickel a bucket for their work. When Professor William Still and I discovered this in 1980, we knew we had a point on which to center our historical research, and I could not resist using it as the basis for the title of the book I began to write in 1982. Even though I got sidetracked by university administrative, teaching, and other research duties, Bill was able to continue refining some of the historical materials into a journal article (Still, Fall 1987).

Since the World War II, shrimping has developed into one of the most profitable fisheries in the United States and, until recently, had become the most important of all fisheries in both the South Atlantic and Gulf

States. Ex-vessel (i.e., catch offloaded from the boats) values number in the hundreds of millions of dollars annually, representing as much as 60% of the value of all fish captured. In North Carolina, the dollar volume has consistently ranged between five and ten million dollars annually, or about 25–30% of the value of all harvested fish combined. However, imports of foreign shrimp, along with other developments, have changed the state's seafood industry dramatically.

Once it became the single most important fishery, North Carolina's shrimp industry became the subject of scientific research, a receptacle for controversy over harvesting practices, the center for competition and conflict among and between various user groups, and a target for more regulation. Excessive harvesting pressure has proven to be one of the most sensitive issues because increased fishing effort results in less economic return per unit of effort; that is, more fishermen are getting smaller individual shares of fish and dollars. Commitments to harvest and market shrimp have promoted some self-sustaining sociocultural patterns that persevere through good and bad economic times. It may be easy to comprehend why more people fish when stock levels are high, but when the prospects for economic returns are low and fishing effort stays constant or even increases, the reasons are less clear. *A main goal of this book, which differentiates it from other treatments of the state's shrimp fishery, is to shed light on this issue (perseverance) through the development of the history and description of the dynamic social processes at work and the seemingly never-ending tensions among them.*[1]

For much of the industry's history, the attention of researchers, resource managers, fishermen, and dealers was focused on the resource, particularly on how the shrimp reproduce, migrate, and grow, and on how they could be most efficiently harvested. Our work in the late seventies and early eighties was the first attempt in North Carolina to understand the social and economic organization of the shrimp fishery—the fishermen, their communities and communication networks, the processing and marketing organizations, and the resource-management structure itself. Earlier, in 1973, a most impressive initial region-wide social-scientific research effort was undertaken under the auspices of the South Carolina Wildlife and Marine Resources Department and the South Atlantic Fishery Management Council (Eldridge and Goldstein, 1975; Calder, et al., 1974).[2]

When we first began our research in the late 1970s, we wanted to better understand how the shrimping industry had become embedded in the larger social and economic structures of commercial fishing in the coastal zone. This is no better illustrated than through first the examination of the annual rounds of fishing and other commercial activities, followed by the nature of the relationship between the fishermen and the dealer/processors, and then on to the distribution network. Work patterns rotate throughout the year in response and as a buffer to changing seasons, stock fluctuations, market conditions, and the sometimes-devastating competition from both domestic and imported seafood. Fish dealers and processors respond by changing purchasing, processing, and marketing patterns. Communities in which fishing and marketing people live adapt as well—sometimes easily, at other times with great difficulty, as we shall see throughout the book.

Obviously these patterns, and changes in them, have important social-policy implications. *This was a second focus our research.* Some of the earliest and most effective efforts to manage shrimping in the southeast United States originated in North Carolina with Ed McCoy and Connell Purvis; their efforts were followed by the work of Dennis Spitsbergen and Mike Street, all of the North Carolina Division of Marine Fisheries. Their tasks involved managing often-conflicting biological, social, and economic goals. And it surely has been fascinating to watch and enlightening to chronicle events as the state of North Carolina has struggled to preserve resources in the face of the countervailing forces at work

Information gathering for this book has involved the examination of public records, such as minutes from meetings, memos, etc.; newspaper articles; published research articles and reports; and personal interviews with people whose history in the fishery dated back to the 1920s. Professor William Still took much of the lead in developing the historical analysis.

Additionally, I used statistical information collected and reported by various fishery-management agencies. John Bort and I conducted several hundred interviews during the seventies and eighties, and I was subsequently involved in literally dozens of research projects where interviewing took place with thousands more people whose responses bear upon this topic in one way or another. A great deal of the community analysis is based on work I engaged in with Paul Tschetter, Barbara Garrity-Blake, and other colleagues. In 2002 and 2003, I revisited dozens of the key informants Bort and I had identified twenty-five years earlier. In some cases I was talking, too, with their sons, nephews, and grandchildren. In others, I have found new informants who have been more than willing to help me bring the information as up to date as possible. I have found, also, a number of recently published manuscripts that bear upon the topics I discuss.

As is the case with all large production systems, the seafood industry generally, and the shrimp fishery in particular, is not without its shady side. I have chosen not to deal with those kinds of matters. Others have, and the results have been informative (Griffith 1993, 1999; Maril, 1983). My purpose is to have readers visualize and enjoy the various kinds of things participants do in going about their work. I must, however, deal with some of the important contextual and internal problems and issues that continually change the nature of the shrimp industry. These are not always that pleasant to deal with for those whose livelihoods depend on shrimp.

I have written this book to reach as wide an audience as possible, including the general public, students, politicians, managers, and, of course, the fishermen and fish dealers. That is why I have tried to present the information in a relaxed, conversational form. If the book does nothing else, it is my hope that those who read it gain a better understanding of the coastal North Carolina fishing culture through the prism of the activities that capture and distribute those tasty creatures we all love to eat—*shrimp*!

John R. Maiolo

Acknowledgements

Beginning in 1979, my East Carolina University colleague, Dr. John Bort, and I began a research project, which was funded by the University of North Carolina Sea Grant Program at North Carolina State University. The project's focus was on the history and sociocultural organization of the state's shrimp fishery. John is an anthropologist. When the project wound down in the early eighties, we began to draft several reports, and I began to put together a book with a title similar to the one you see here.[1] We also enlisted the aid of Dr. William Still, the university's nationally known maritime historian, to help us develop some of the narrative on the history of the fishery and its regulatory framework. Other duties got in our way, and, as a team, we were never able to finish the book. John Bort went on to other academic pursuits. Bill Still retired. I did, as well, in 2000.

I never could accept the fact that this book would not get finished. Over nearly a quarter of a century, I continued with my research on the shrimp fishery whenever I had the chance. As time passed, it became clear the early studies would serve as a solid baseline against which to measure changes. So, after retiring, I, with several colleagues, first finished my work on the volume we had undertaken on the aftermath of Hurricane Floyd (Maiolo, et al., 2001). This had to be done right away in order to catalogue the critical issues surrounding that disaster. I then pulled out the early 1980s drafts of the book, and began to piece together the team's and my notes and fieldwork, past and present. Neither John Bort nor William Still was available to return to the project, so I took it on myself. Throughout the book, there are places where they contributed a great deal to the early narratives. I try to make this abundantly clear where it is appropriate.

Marcus Hepburn helped me connect with fishermen, especially the channel-netters who are the subject of Chapter Three. In the early phases of the research, I hired his spouse, Toni, to create some of the wonderful handcrafted drawings of the shrimp boats and channel-net activities you will see throughout the book. Kevin Fontana and B. J. Reckman assisted in the preparation of the images I use in the book.

I wish to thank current and former staff members of the North Carolina Division of Marine Fisheries (NCDMF)—Manley Gaskill, Jack Guthrie, Ed McCoy, Bob Pittman, Connell Purvis, Dennis Spitsbergen, Mike Street, "Red" Munden, Preston Pate, and Katy West. Nancy Fish, Sean McKenna, Brian Cheuvront, Janice Fulcher, Chris Wilson, Dee Lupton, Chip Collier, Rich Carpenter, Cheryl DeBerardinis, and Doug Mumford, also of the NCDMF, were extremely helpful in my being able to keep the research materials updated. Bill Reeves of the New Hanover County Museum was helpful in identifying historical materials during our research in the 1980s. Dr. B.J. Copeland, director of Sea Grant at the time we received the research grant, was always supportive. So was Bob Hines, also with Sea Grant. I am grateful to Dr. William Queen and Cindy Harper of the Institute for Coastal and Marine Resources at East Carolina University for their decades of support and assistance.

The fishermen and fish dealers of North Carolina were the subjects of our research, and their patience and cooperation was, and still is, outstanding. They lead an exciting but difficult existence. It is to their well-being that this book is directed as a token of gratitude. I hope that somehow, it will make the world around them a bit easier with which to cope and to understand. I want to extend a special thanks to Mr. Clayton Fulcher. Probably more than anyone, he informed me of the value of trust and long-term relationships in the seafood industry. His grandson, Tommy, informed me of the key changes in the industry during the recently past decades, and provided valuable pictures and documents from his personal collection for this book. Doug Brady and Jule Wheatley have answered every phone call and every question I ever posed to them (for the past twenty-eight years)! What more can one ask of busy people?

I want to express my thanks to Andy Scott, a publisher and friend, who made valuable comments on one of the early drafts of the book. Lee Maril's book, *Texas Shrimpers*, served as a constant source of inspiration to finish this project.

I am deeply grateful to Dr. Tom Feldbush, Vice Chancellor for Research, Economic Development, and Community Engagement, and Andrea Harrell, his Assistant Vice Chancellor, who have, through their good offices, provided valuable financial assistance for this volume and our earlier book on Hurricane Floyd.

Karen Willis Amspacher, of the Core Sound Waterfowl Museum, was incredibly helpful in my obtaining crucial information on Harkers Island, including recently published materials that might have otherwise escaped my attention.

Of course, the book you hold would not have come to be without the professional expertise of The Chapel Hill Press: Edwina Woodbury (publisher), Misty Thebeau (publisher's assistant), Katie Severa (book and cover designer), and Jeff Thomas (editor). My thanks to each of them for making this one of the easiest, most pleasant publishing experiences of my career.

Finally, I want thank my wife, Ruth, for her unwavering support for my research and her understanding of the costs, both financial and human, associated with an academic career emphasizing research and publication. It is to her that this book is dedicated.

COASTAL NORTH CAROLINA COUNTIES

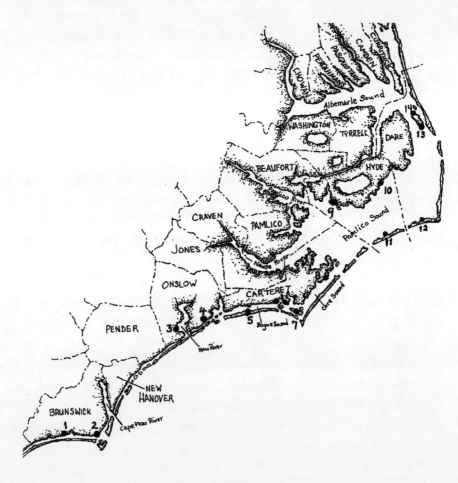

PORTS

1. Holden Beach
2. Southport
3. Sneads Ferry
4. Swansboro
5. Salter Path
6. Beaufort/Morehead
7. Harkers Island
8. Atlantic
9. Swan Quarter
10. Englehard
11. Ocracoke
12. Hatteras
13. Wanchese
14. Manteo

Introduction to the Fishery

As an introduction to the shrimp fishery, it is important to understand the nature of the creature itself. Because of differences in how, when, and where they spawn, migrate, and grow and what kinds of factors influence these events, the various types of shrimp North Carolina fishermen pursue impose the *initial* set of constraints on, as well as harvesting opportunities for, the fishermen. The related social, cultural, economic, and management structures provide the others.

I do not, in any way, intend this to be anything more than an ever-so-brief summary of research on the reproduction, migration, and habitat of the shrimp. Those who wish to know more than I provide here can consult what I still believe to be an extremely understandable and well-researched report, *The Shrimp Fishery of the South Atlantic States: A Management Planning Profile*, produced in 1974 for the South Atlantic Fishery Management Council (SAFMC) by the South Carolina Marine Resources Center in Charleston. The SAFMC is one of eight regional councils established under the Magnuson Fishery Conservation and Management Act of 1976 and is charged with developing management plans for fisheries for waters extending from three miles to 200 miles seaward along the Southeast coast. Its jurisdiction encompasses North and South Carolina, Georgia, and the east coast of Florida.

THE SHRIMP'S LIFE HISTORY AND OTHER BIOLOGICAL STUFF Commercial shrimping in North Carolina focuses exclusively on three species of the

Penaeid family. They account for all but a modest amount of the shrimp yield throughout the Southeast Atlantic and Gulf of Mexico regions. Florida (and, to some extent, Georgia and even South Carolina) can experience commercially significant yields of rock shrimp, which biologists refer to as *Sicyonia brevirostris*. Since separate records have been kept, yields in the South Atlantic have been quite small, ranging between 1.86 and 22.18 million pounds. The average for the decade of the nineties was 6.12 million pounds per year. During the same period, North Carolina produced an average of only 5,800 pounds and, in some years, yields in only the hundreds of pounds. A fourth, deepwater shrimp, popularly referred to as the *royal red*, is found in limited quantities off of Florida's east coast.

On average, two thirds of North Carolina's yield consists of the *Penaeus azteca azteca* species, popularly known as the brown shrimp, brownies, green lake, or summer shrimp. *Penaeus setiferus setiferus* is known as the white shrimp or green tail because the tips of its tail fans (technically known as the telsoms) display an iridescent greenish hue. *Penaeus duorarum duorarum* is most popularly known as the pink or spotted shrimp. To a limited extent, pink shrimp are found near Bermuda; whites as far north as Fire Island, New York; and browns north to Martha's Vineyard. But estuarine and ocean abundance has not warranted the development of commercial shrimping north of Cape Hatteras.

Most consumers do not know the differences among the three *penaeids* when they buy them, nor do they care. When shrimp are iced or cooked, differences in coloring are imperceptible to the untrained eye, and there is absolutely no difference in taste. However, a notable exception in market preference has been displayed by Chinese Americans, especially restaurateurs, who, until recently, have preferred the *P. setiferus* when provided a choice.

How and when shrimp reproduce is of considerable importance to the managers and producers because there has been a remarkable increase in harvesting pressure from full-time, part-time, and recreational fishermen as the popularity and price of shrimp have grown. Shrimp mating habits are not well understood, but, apparently, mating among whites occurs only between hard-shelled animals, not during shedding. Shedding occurs regularly as the animals grow. New, larger soft shells are already present when the older, hard shells are shed. The new shells harden quickly.

For browns and pinks, copulation occurs between hard-shelled males and soft-shelled (shedding) females (Perez Farfante, 1969). There is no evidence of romance or foreplay. Fertilized eggs 1/75 of an inch in diameter are then laid in the water. They are *demersal* (bottom dwellers). They hatch in about twenty-four hours.

All three of the species spawn offshore. Water temperature is critical. Whites spawn in the spring and fall in relatively shallow water, near inlets close to the shoreline in depths of twenty to eighty feet. Browns spawn during the fall, winter, or early spring, depending upon geographic location and water temperature. Pinks spawn during the spring, summer, and fall. In North Carolina, the peak spawning periods are the spring for browns and summer for pinks (McCoy, 1972).

Shrimp are great hitchhikers, and nature provides currents to carry them to different locations for mating, spawning, and growing until they become juveniles and then adults. In the last two stages they have the means for *active* locomotion through crawling, swimming, or propelling themselves using their tail section. But, in the meantime, after spawning, friendly water flows carry the larvae farther seaward for safety as they pass through no fewer than eleven larval and post-larval stages. The transition takes from ten to twenty-five days, depending upon the species and whether water temperatures, food, and habitat conditions are suitable.

After over-wintering in deep, warm, ocean waters (with some exceptions, which I will note below), post-larval migration toward the estuaries begins. In the early spring, brown post-larval shrimp, which spawned in the *previous* fall, emerge from burrows in the ocean floor first, followed by the migration of white and pink shrimp post-larvae in late spring and summer. The estuarine bays, sounds, and saltwater rivers provide rich feeding grounds and protection from natural predators.

Adolescent shrimp can grow as rapidly as one to two-and-a-half inches per month with favorable habitat conditions. Under similar conditions, browns grow the fastest, followed by the whites and pinks. Research has revealed that average shrimp sizes are larger when the population density is lower (see reports by the South Carolina Marine Resources Center and the SAFMC).

Figure 1:1 gives a general reference for spawning and growth cycles in North Carolina by location and time of year. The reader is reminded that North Carolina is the northernmost boundary for commercially viable quantities of *penaeids*. Figure 1:2 diagrams the lifecycle of white shrimp to provide an idea of the migratory flow of that species.

All three species are omnivorous and feed on a variety of organic materials and small organisms, mainly at night. These behaviors are important to commercial fishermen and affect the shrimpers' choice of various types of gear to maximize catches in different water conditions and at different times of day.

Salinity levels in nursery areas are crucial to the abundance of surviving shrimp crops. They are particularly crucial in April and May, after juvenile browns have already migrated into the upper reaches of the estuaries for further growth. Pinks prefer higher rates of salinity than browns, which accounts for differences in estuarine locations among the species. Browns and whites are found farther upstream in the lower salinity, brackish waters.

Catches, or "landings," are affected greatly by both natural and human made events, which affect water temperatures and salinity. For example, during the 1977–78 season, in addition to experiencing an unusually cold winter and spring, North Carolina's coastal zone experienced heavy rainfall in April, and commercial landings dropped to their lowest level in twenty years!

Shrimp Species Variations in the Biological Life Cycles in North Carolina[1]

	Jan	Feb	Mar	Apr	May	June	July	Aug	Sept	Oct	Nov	Dec
W H I T E					Spawning		In Nursery		Leave Nursery Peak Catch			
B R O W N			Spawning		In Nursery		Leave Nursery Peak Catch					
P I N K					Peak Catch			Spawning	In Nursery		Leave Nursery	

FIGURE 1.1   [1]*Source: Originally adapted from G. M. Knopf,* Opportunities in the Shrimp Fishery Industry of the Southeastern United States. *SEA GRANT Information Bulletin No. 3, January 1970: 16. I thank David Whitaker of the Division of Marine Resources, South Carolina Wildlife and Marine Resources Department; and Dennis Spitzbergen and Sean McKenna of the NC Division of Marine Fisheries, for advice in the modification of the model for North Carolina.*

Shrimp is an important food item for many marine species. Other shrimp, crabs and some thirty-four species of finfish rely on the *penaeids* for sustenance. Marine predation can reduce the potential harvests by as much as two-thirds. Humans are also predators, using shrimp not only as food, but also as bait for sportfishing.

HUMAN IMPACTS AND SOLUTIONS   Humans impact the shrimp population in more ways than capturing and eating them or using them as bait. First, coastal farming and development related to tourism produce increased runoff of freshwater, silt, and chemicals, affecting levels of salinity, water column and bottom quality, water temperature, and so on. Second, commercial and recreational fishing activities themselves can seriously damage

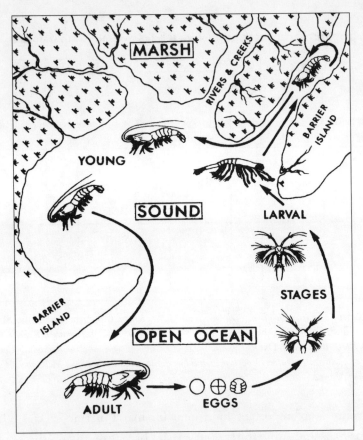

FIGURE 1.2 *Diagram of the life cycle of the white shrimp (cycle occurs in a 12-month period). Source: ASMFC October 1965.*

or even destroy habitats. For example, in North Carolina, an important estuarine fishery, especially during the winter months, is clamming by mechanical means. The ever-resourceful fishermen in the state adopted a method of harvesting called "clam-kicking" in 1977. This was based on a prototype method used on small, twenty-five-foot skiffs in the 1940s. But, in the seventies, vessels up to sixty feet were outfitted with a plate that creates a downward thrust of the engine wash. The turbulence disturbs the muddy bottom so that a following dredge, hooked to the same boat, can skim the dislodged clams. Proponents of the method argued for the efficiency of the technique; opponents rightfully argued that the method killed over-wintering shrimp either by destroying their habitats or by stirring up particles that get trapped in their gills, resulting in suffocation. The animals can also be washed out of the protective bottom silt and killed directly by the cold winter water. The practice has continued to this day, but the areas where it is allowed have been severely restricted.

Dredging for scallops poses a similar threat, as do crab trawling and shrimping itself. Depending upon the month of the year, habitats can be damaged or juveniles too small for market can be captured, wasting the resource.

Protection of shrimp and finfish habitats has received a great deal of attention since the 1960s. In 1992, the state began to take even more aggressive preventive action, which expanded into a fairly robust set of rules by the turn of the century to protect a full range of marine species, not just shrimp (NCDMF, March 1999). Marine resource managers responded to the problem by restricting clam-kicking and scallop dredging, limiting *when* and *where* those activities can occur.

Shrimp and crab trawling have long been severely restricted in designated nursery areas. Seasonal restrictions have been placed on these activities, as well. When management of the fishery became grounded in scientific research during the 1960s (discussed in detail in Chapter Five), an emphasis was placed on the designation and protection of nurseries for all species. Salt-marsh estuaries serve as nurseries for 90% of the state's important fisheries. North Carolina became the first state to formally protect these fragile ecosystems. Three categories were designated: primary, secondary, and special secondary nurseries.

Primary nurseries are those in the upper creeks and bays. These nurseries consist of shallow, soft, muddy bottoms, with surrounding marshes and wetlands. Their low salinity levels and abundance of food are ideal conditions for both finfish and shellfish. Along with inlets, they are recruitment bottlenecks for many economically important species. Many commercial activities—including trawls, and seines—are prohibited throughout the year in these areas. Prohibitions also exist for dredges and any other mechanical means of harvesting clams or oysters. Currently, there are 80,144 acres under this designation along North Carolina's coast.

Secondary nurseries are in the lower portions of creeks and bays. Young fish and shellfish move into these waters after leaving the primary nurseries. No trawling is allowed in these areas, consisting of 35,502 acres. There are also *special secondary* areas, usually adjacent to the designated secondary nurseries, but closer to the open waters of the sounds and ocean. These areas are closed only when there is proof of an abundance of juveniles during certain periods of the year. The special secondary areas consist of 31,362 acres. When appropriate, secondary nurseries are opened for trawling when either the shrimp are believed to be of commercial size or the seaward migration begins.

Needless to say, a great deal of controversy remains between managers and fishermen over when and where shrimping should be allowed. Fishery scientists and resource managers want to protect the habitats where the post-larval shrimp migrate either to grow or to over-winter. Operators of large boats want more restrictions on estuarine shrimping to allow as much of the resource as possible to migrate to deeper waters where *they* can capture them. Operators of small boats, many of whom are part-time-commercial or recreational fishermen, want access to the shrimp early and often (some say always and everywhere). The resource managers somehow cope with this controversy through a delicate system of communication, involvement in test trawls and public hearings, and personal persuasion. As we shall see later, confrontations among the user groups often occur.

Because of the growing pressure on many marine species throughout the coastal waters of the United States, generally, and North Carolina, specifically, a great deal of resource-management attention has been directed toward the question of overfishing, of which there are two distinct types. *Recruitment* overfishing is a biological question, and occurs when stocks are harvested to the point at which the population of a species declines due to curtailed reproductive activity. This occurs when harvesting leaves too few sexually mature adults available to maintain the stock at prior levels and/or when sexually premature fish are over harvested, which prevents the continued replacement of mature males and females.

There is some controversy over the issue of recruitment overfishing in the shrimp industry. Historically, it has been thought that this could not occur in the shrimp fishery because it could only happen in fisheries of finned species, such as trout, flounder, mackerel, bluefish, etc. that have a relatively limited reproductive capacity. A single female shrimp can produce an astonishing number of eggs—as many as 700,000 or more per breeding season! Some managers and scientists argue that enough eggs are produced by a small number of egg-bearing females to easily replenish the stock. Counterarguments involve migration, spawning, habitat, habitat destruction, predation, and climatological conditions that I have already discussed.

There has been some concern over whether white shrimp can, indeed, be over-harvested in the waters off South Carolina, Georgia, and the east coast of Florida. This concern is based on the heavy fishing pressure during the migration of whites from Florida northward. It has been suggested that weather conditions, in combination with heavy harvest pressure, can reduce reproductive capability and, therefore, stock levels. Evidence to prove the assertion is still a source of controversy, but federal regulations were put in place in 1993 (SAFMC, June 1993). The new rules specify conditions under which white shrimp fishing is prohibited, mainly in Georgia and South Carolina. North Carolinians have complained that the rules are really intended to keep Tar Heel fishermen out of those states' waters and to protect the local fishermen rather than the fish.

Others acknowledge a reduction in overall yield, as measured by the *number of pounds landed*, but the cause is believed to be *growth*, not *recruitment*, overfishing. Growth overfishing is an economic issue and occurs when fish that do not produce the highest attainable economic yield are harvested, even though the reproductive capability of the fishery is not threatened. Growth overfishing always precedes recruitment overfishing. That is, the fishing pressure first captures the largest and most available of the species, which produce the best economic yield per pound. Their availability presents the least cost for harvesting. As the size is fished down, harvesting pressure seeks the remainder of the stock, and prices drop for the small animals. Eventually, this trend can result in recruitment overfishing.

In most cases, when shrimp are sold from the boat to the dealer, they are sorted and graded by size, and the price differences among various sizes can be substantial. Because of this, shrimpers and dealers speak in terms of count per pound (i.e., how many creatures it takes to make a pound). In these terms, the fewer, the better; for example, 20-count shrimp might well be used in a restaurant, while 60-count animals would generally be perceived only as baitfish. Shoppers can see this in fish markets that carry various sizes, although most markets simply classify the product as large, medium, or small. It is also important to note that counts are related to whether the animals are headed (more properly, *be*headed) and whether they have been peeled; the larger shrimp lose a lower percentage of their total weight through heading and peeling. Shrimp can lose up to half of their body weight when headed, peeled, and de-veined. So a thirty-count batch of headed and peeled shrimp would cost almost twice as much as 30-count with heads on. That is why the mid-size shrimp (20–26 count) are the most popular. The largest ones (12–16 count) are the most expensive. The type and length of cooking for which they are suited are also related to size.

There are many regulations that restrict harvesting until the shrimp reach counts considered to help optimize economic yields, and this is important for the public to understand. Yield, then, may not always be a consequence of how many animals were harvested, but how large they were at the time of capture. But resource managers also have to be aware of migratory patterns. When migration begins, even if the animals are smaller than managers would like, they must be harvested or they may be lost to the commercial fishery. Shrimp are considered to be an "annual crop" because their lifespan is about a year (some whites have been found to survive until a second season), so they must be harvested soon after reaching maturity. When shrimp reach larger bodies of water, they spread out, are not sufficiently concentrated to make commercial fishing profitable, and can be lost to the fishery. White shrimp are an exception here, as well, because they stay congregated and continue their migration close to the shoreline (for example, in the Atlantic Ocean, from Cape Canaveral to southern North Carolina). Thus, except for the whites, once outside the inlets, the prospects for capture diminish sharply. It is interesting to note, too, that because whites tend to migrate close to the shoreline and across state lines, some managers feel it is quite appropriate to harvest them, even if they are smaller than the economic optimum, *before* they reach the next state and their value to local fishermen and state statistical totals are lost.

There is little controversy among the resource managers and the fishermen about the constant threat of growth overfishing, even if there may be disagreements over specific policies (exceptions are discussed in Chapters Four and Five). For that reason, all of the states in the South Atlantic and Gulf regions have management policies to govern shrimp harvesting in a way that produces high yields. North Carolina's policies deal with area and seasonal restrictions, as we noted, that function to protect habitats, as well. At a broader level, the Gulf of Mexico Fishery Management Council some years ago placed into effect a plan to govern shrimp fishing in the Gulf out to two hundred miles. The South Atlantic council considered but did not adopt a similar plan for North and South Carolina, Georgia, and the east coast of Florida.

The term "bycatch" (also referred to as "incidental catch") refers to harvesting one kind of marine species while in the act of targeting another. The Atlantic States Marine Fisheries Commission has further refined the definition to refer to the "non-selectivity of the fishing gear to either species or size differences" (NCDMF, 1999:9). In the case of shrimping, a number of serious issues have been raised during the past several decades that have created controversy and management action.

One hotly contested battle occurred in the 1970s over the killing of various turtle species; such bycatch kills numbered in the thousands in the South Atlantic states, including North Carolina. Anecdotal information was readily available in the sixties and seventies linking the killing of sea turtles to shrimp trawlers. Then, in the late seventies, the National Marine Fisheries Service (NMFS) received a report of unprecedented numbers of turtles being caught near Cape Canaveral (SAFMC, 1993). Studies were initiated that confirmed the fears of massive turtle mortality resulting from shrimping. While these studies were underway, NMFS initiated a program to remedy the situation. Various experiments with tow times—that is, the length of time towing for shrimp could occur—failed to stem the destruction. As part of the search for a solution, Turtle Excluder Devices (TEDs) were developed to attach to trawl gear. These devices were placed into part of the shrimp trawl net to allow turtles to escape while the towing continued.

Bycatch of other marine species again became an issue when scientists reported that shrimp trawls were found to be responsible for more discards of incidental catches of non-targeted fish than any other form of commercial fishing. I say "again" because, as we shall see in a later chapter, bycatch from trawling had been an issue of great concern for decades.

However, not all are in agreement that these bycatch estimates are all that accurate (NCDMF, 1999:10). In 1995 NMFS developed a set of standards that enhanced confidence in estimates. Since then, the NCDMF has concluded that four species of finfish, already stressed by targeted commercial and recreational fishing, were being further depleted to dangerous levels because of shrimp-trawl bycatch. They are weakfish (sea trout), croaker, spot, and summer flounder. Research indicated that incidental catches removed spawning stocks of these species, in addition to fishing down the abundance (NCDMF, 1999: 7–20). We will revisit this issue again in Chapter Five.

The use of TEDs to prevent turtle mortality became a decade-long drama that included trial regulations, lawsuits, temporary injunctions, emergency rules, suspensions of the rules, and revisions of regulations, which were at last finalized in the nineties. The South Atlantic Fishery Management Council (1993) has published *four pages* chronicling the agency and legal actions on the journey to final resolution: TEDs must be used when trawling for shrimp with few exceptions, mainly having to do with the type of gear used.

With the expanding concern for other types of bycatch, which could not always be dealt with by TEDs, managers and the governments turned their attention to the development of BRDs (Bycatch Reduction Devices). In 1995, the South Atlantic Council adopted an amendment to the Shrimp Fishery Management Plan requiring the use of BRDs by shrimp trawlers in federally controlled waters. In 1992, the NCDMF issued a proclamation that requires one of four approved BRD designs in all shrimp trawlers (SAFMC, 1999:41; also, see Chapter Five of this book for further discussion of this topic). Adequate stock assessments are not yet available to determine the effect on all stocks, but the NCDMF has reported an improvement in the weakfish population.

Let us now turn to an examination of the historical, social, cultural, and economic features of the North Carolina shrimp fishery.

From Leafy Weirs to Otter Trawls

(WITH WILLIAM STILL[1])

Historically, a wide variety of craft, gear, and fishing techniques has been and continues to be used in shrimp harvesting. This is not simply a reflection of advances in capture technologies. Shrimping, from harvest to market, has become embedded in the cultural patterns of the region, and the continuity of the industry can be most effectively understood in these terms. This approach ignores neither the persistent winds of change nor influences from outside the fishery and the region. Instead, we seek to better understand how the forces of change, such as technology, are managed within the fishing community and occupational networks. Further, we concentrate on the selective adoption of new ideas and technologies. In this chapter, we focus on the development of the fishery into the 1930s. It was particularly during the period after 1914 that many of the cultural forces emerged to give the industry much of the shape that has endured into the twenty-first century.

BEFORE THE WAR BETWEEN THE STATES Native Americans caught shrimp for subsistence using dip nets, seines, and weirs (traps) made from leaves and cord (made of fibers possibly extracted from bear grass and Palmettos). Early European settlers in North Carolina captured shrimp in a similar manner for local consumption. John Lawson, who traveled extensively in the colony during the early eighteenth century, wrote in a book first published in 1709, "shrimps are here very plentiful and good, and are to be taken with a small Bow-Net, in great quantities." A gentleman by the name of John Brickell wrote in the 1737 edition of the *Natural History of North Carolina* that shrimp were "very restorative and good in Consumptions, Hecticks, and Asthmas" (also quoted in Still, 1987:257).

Up to the Civil War, fisheries in Carolina were mostly of local importance. The availability of fresh seafood was restricted almost entirely to the isolated coastal communities. There was a limited trade in salt-treated fish, and some shellfish trade was established with the state's interior and the northeast. At this time, the most important fisheries were shad, sturgeon, mullet, oysters, striped bass, alewives, and drum.

There is some indication in the state's fishery-management literature that shrimp from North Carolina were transported to Philadelphia during the eighteenth century. However, there has been no indication about how the twin problems of preservation and transportation were solved, since this occurred before railroads and steamboats were available (Joyce and Eldred, 1968:38).

AFTER THE CIVIL WAR In the 1880s, R.E. Earll, upon investigating shrimping in North Carolina, reported, "The fishermen eat very little themselves and throw the bulk of their catch away" (1887). There is, however, evidence that some commercial interest was beginning to develop even as Mr. Earll was conducting his research. He seemed to recognize this by noting that about fifty fishermen had begun to direct some of their efforts toward shrimping before 1880 in Wilmington. Actually, it seems that the harvesting activity was really taking place southeast of there in Southport. Newspapers in the area began to mention the local consumption of whole shrimp "boiled and peeled." During this same period, pickled shrimp became common in local fish markets, were sold in quart and gallon containers, and some were shipped out of state (both fresh and pickled).

Until the development of refrigeration, shrimp, due to its extremely perishable nature, was severely limited as a marketable commodity in areas not near the locations of harvesting and off-loading. Although ice had been used in the United States as a preservative for fish since the 1830s, it was not used extensively as a refrigerant for fish until the 1870s. In the early seventies, George Ives, originally of New Haven, Connecticut, introduced the iced-fish trade into North Carolina. For a number of years, however, ice continued to be a scarce commodity, most of it having to be shipped in during the wintertime. In 1889 there were only five ice plants in the state, which increased to twenty-three by the turn of the century. Ice continued to be shipped in from the northeast during the winter and was occasionally obtained when local streams froze over. The New Bern *Daily Journal* (January 13, 1886) noted that the Trent River was frozen and "the fish dealers were busy ... packing ice for future use ...." The Wilmington *Morning Star* reported (November 14, 1884) that the North Carolina fisheries had become the most important on the South Atlantic coast. However, according to the same paper, in the coastal areas south of the Albemarle Sound they were practically undeveloped because of the lack of shipping and "refrigerating conveniences."

Inadequate transportation was another factor in retarding the development of markets. Although Morehead City and Wilmington were served by rail transportation before the Civil War, the small fishing villages and coastal littorals were not. Southport, which would become the center of the shrimping industry for many years, would not be reached by rail until 1912. Although boat transportation was used to transport fish products, there was no refrigeration. The same was true of the railroads (Wilmington *Morning Star,* May 31, 1912; Carteret County News Times, October 11, 1955).

Inadequate vessels and inefficient harvesting gear were also detrimental to the development of a commercial shrimping industry. Before the twentieth century, the only powerboats used in North Carolina fisheries were small steam flats for harvesting shad in Albemarle Sound. Small rowing and sailing vessels were most common in the state.

California was the first state to consider shrimping to be important commercially. In 1869, eight boats manned by Italian fishermen were engaged in the fishery, using small-meshed seines. Two years later, Chinese immigrants began shrimping and shipped large quantities of dried product back to China. In 1880, 1.2 million pounds, valued at $124,000, were landed in California, far more than in any other state, up to that period (Earll, 1887: 803, 810; Johnson and Lindner, 1934:63). In contrast, 5,000 bushels (about 25 to 40 thousand pounds) were landed in the Wilmington area, but only half were sold for food, nearly all locally. The remainder was sold as bait and fertilizer (Earll, 1887:802–3).

VENTURE CAPITALISM During the first decade of the twentieth century, the South Atlantic and Gulf States became the center of the shrimp industry in the nation. In 1908 the catch in the region was slightly over 18 million pounds, which represented 95% of the total catch nationally. During these years, the industry in North Carolina made steady but slow progress, still being confined primarily to the southeastern corner of the state. The *Southport News*, in October 1914, reported:

> The fish business has been the shrimp, or to be correct, the prawn has and continues to engage the attention of all local toilers of the sea. The way was shown how the prawn could be scooped up by net, and at once the fishermen prepared and went after the prawn or "big shrimp" going along the coast .... The prawn industry has proven profitable all around and given employment with good wages while the fishermen have taken for the times big money as can be seen when a boat's crew catches anywhere from ten to fifty bushels and receives from $1.10 to $1.30 a bushel. Shrimp catching has been so profitable that other kinds of boat fishing have been passed by, the results being that the local market has been fare of fish.

At about the same time, it was generally believed in Southport that if one heard the plea "hard times," the answer would be a resounding, "shrimp," which meant the harvests were good. Many in the area seem to have been astounded by the fact that an abundance of shrimp lay in nearby coastal waters, and it took until about 1914 for the industry began to develop.

Previously, Wilmington's *Morning Star* had noted (September 27, 1912) that the completion of the Wilmington, New Brunswick, and Southern Railroad to Southport had resulted in the establishment of a firm to pack and ship fresh seafood by railway express. By 1914 shrimp (headed and iced) were being sent to northern states by rail. The *News* reported a shipment of 300 bushels to New York City by refrigerator car.

Although the state's southeastern counties dominated the commercial shrimp industry during the first decades of the twentieth century, some shrimp was shipped to eastern markets from farther north in the state, specifically in the Carteret County area. One article in the *Atlantic Fisherman* noted, "the inception of the commercial shrimp industry (in the southern states) is often connected with Charles S. Wallace who, in 1907 first shipped shrimp from North Carolina to New York" (September 1953). Mr. Wallace was a fish dealer located in Morehead City, Carteret County. As we shall see, however, the commercial shrimp fishery did not become a major harvesting activity until the 1930s.

Two developments stimulated a dramatic growth in the industry just prior to the outbreak of World War I. First, a group of Southport businessmen contacted fish dealers in a number of northern cities and, on the basis of the encouragement they received, decided to build a shrimp-canning factory in their community. The first such cannery had been established in New Orleans in 1867. By the summer of 1915, the Southport cannery was up and running and was so successful (employing over fifteen people and producing more than ten thousand cans a day) that a Massachusetts businessman established a second one in nearby

Shallote. By enabling large quantities of shrimp to be marketed more easily outside the region without the worry of spoilage, these plants made commercial shrimping feasible. A third plant later was established in Southport in the 1920s, as reported in the Carteret County *News Times* (October 11, 1955). This portion of the industry, however, experienced a total collapse during the Great Depression.

Second, from the standpoint of harvesting, the adoption of the otter trawl enabled the capture side of the industry to keep pace with market growth (for a detailed description of otter trawls and their use, see McKenzie, 1974:43–48). The otter trawl has been called the greatest of all fishing-gear innovations. The development of the powerboat at the end of the nineteenth century, and its rapid adoption by commercial fishermen, made the use of otter trawls practical. One source indicates that the trawl was developed from the English beam trawl and introduced into this country in New England during the 1890s (Jensen, 1967:6).

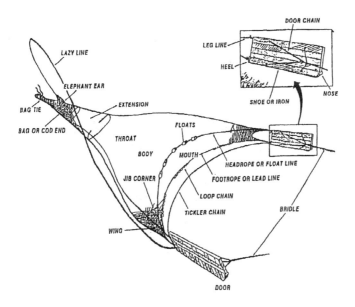

Sketch of a typical otter trawl and components. Adapted from NMFS 1992.

Chapter Two

Another asserts that the otter trawl was first made in Scotland in 1892 (DMF, March 1999:4).

The otter trawl revolutionized shrimp fishing. A long-haul seine net could be used only in shallow waters and required a large crew to purse the seine. The efficient otter trawl could be used either inshore or in deep water, needed few men to operate it, and yielded a greater production per unit of effort.

Otter trawls range in length from about thirty up to fifty feet. They are attached to vessels by two lines extending to the trawl "doors," which are still made from oak and are from two to three feet by four to ten feet in size. These doors spread the opening of the net to herd the fish into the tail, or "cod end," of the net. A lead line extends from the doors at the net's bottom, while a cork line is attached to the top. The floats and weights determine the shape of the net as it is pulled through the water. Fewer floats are used to create a "flat" net (that is, more rectangular) for capturing brown shrimp that have burrowed into the bottom after being frightened by the net moving toward them. A standard "tickler chain"

Sketch of an ocean going shrimp trawler.

A skimmer trawl.

(also attached to the doors) is dragged ahead of the net to frighten the shrimp off of the bottom.

Over the years, various gear modifications have developed to deal with different conditions and improve harvest capability, but the basic concept of the otter trawl has remained intact. The "skimmer" net was developed to fish for shrimp in the sounds and rivers. This is a different technology than the skim gear discussed earlier. The mouth of the net is kept open by "frames" off the sides, while the bunt, or cod end, is near the stern of the boat. While the net is out, the vessel is turned into the tide and uses just enough power to hold her steady while the tide does all of the work. This, in a way, is channel-netting with power and movement. Channel-netting is a very special type of fishing and is the subject of the next chapter.

One processor/dealer told me during an interview in the winter of 2003 that the skimmer is much more efficient than other nets in shallower water. He noted that an older shrimper using a standard net with a tickler might catch three hundred pounds in a night's work. Right next to him may be a younger and more adventurous shrimper with a skimmer who would have caught a thousand pounds the same night.

In contrast to flat shrimping nets, "balloon" and "semi-balloon" nets are used for white shrimp—which, over the years, were discovered to be able to jump over the net—or for either of the other species that might swimming at mid-depth or near the surface. The largest boats use twin nets: that is, they will pull two large nets side by side and with a device in between to prevent tangles. Some will use "four barrels," two sets of twins, which use less power and have less drag. I will return to a discussion of this later.

"Try-nets" complete the array derived from otter-trawl technology. These are miniatures of the larger trawls, being normally only sixteen feet wide, and are used in two ways. Larger vessels use them to determine if it is worth the trouble to set the larger nets. Smaller vessels use them as the main trawl for shrimping in the estuaries.

Exactly when American fishermen first adopted otter trawls has not been firmly established. Between 1912 and 1914 the U.S. Bureau of Fisheries, the forerunner of the National Marine Fisheries Service (NMFS), used a small version of the otter trawl for collecting marine specimens at its Beaufort, North Carolina, center, and it was reported that large quantities of shrimp (possibly as bycatch) were taken during the hauls. In 1913, a group of Portuguese and some Italian immigrants had used otter trawls for shrimping near Fernadina Beach, Florida (Johnson and Lindner, 1934:9; Captiva, 1967:24). There is no evidence, however, that the two uses were connected. Then the Charlotte *Daily Observer* (June 7, 1915) reported that, in 1914, *Norwegian* fishermen from Fernadina Beach used an otter trawl in North Carolina waters near Southport. In March 1916 the Wilmington *Dispatch* noted that it was a young New Jersey fisherman, Samuel Thompson, "who brought the first deep-sea shrimp net to Southport ... and taught the ... fishermen to catch shrimp." In September 1916 the Wilmington *Morning Star* mentioned the use of "large drag nets" to harvest shrimp in the ocean. The newspaper estimated that there were fifty boats using the gear by the end of the year. Other types of gear—such as hand seines, cast and push nets, and even pound nets—were still in use after 1920, but trawls rapidly became the primary gear for shrimping. From these news accounts, it is apparent that the otter trawl was rapidly adopted in North Carolina once the gear became known. By 1927 the otter trawl was in widespread use throughout the South Atlantic and Gulf regions. Between 1908 and 1918, ex-vessel yield increased from 18 to 48 million pounds! By 1945, the amount of shrimp caught with otter trawls had increased to over 95% of the total catch (Chestnut and Davis, 1975:297 passim).

Until the adoption of otter trawls, many shrimpers operated without boats, since most of the harvesting was done in shoal waters close to the shoreline. Those boats that were in use varied in size and shape—for example, sailing vessels, rowboats. With the appearance of the otter trawl, vessels specifically designed for shrimping were built and brought to the fishery. The earliest trawlers were open skiffs, from fifteen to twenty feet long, powered by small gasoline engines (Captiva, 1967:24). By the early twenties, the first "decked" trawlers were in use, and that design still prevails. It was developed from the so-called "Mediterranean work boat," similar to the Greek vessels used in the sponge fishery off of the coast of Florida (McKenzie, 1974:40).

The classic ocean-going wooden vessels that you see docked at the ports in the state's coastal villages are in the thirty-five- to seventy-five-foot range. Converted automobile engines often power the smaller boats used for fishing inside the inlets. Larger ones use diesels. Some are pushed by single, others by twin engines. The pilothouse is forward, with the galley and sleeping quarters behind it. At the rear of the vessel are the winches and outriggers that place the nylon trawls in the fishing position or bring the nets up for either dumping the catch on deck or storing the nets in a non-fishing position.

Chapter Two

FROM WORLD WAR I TO THE DEPRESSION World War I had very little effect on the development of the state's shrimping industry. On January 10, 1918, the U.S. Food Administration, which controlled the fishing industry during the war, announced that "all saltwater fishermen ... whether fishing independently or in shares, engaged at any period of the year, in the commercial distribution, including menhaden and shellfish and crustaceans," were required to secure licenses. In April the same authority removed all state initiatives to control fishing that were "not absolutely necessary for the conservation of the fish supply." These actions may have been taken to stimulate commercial fishing. Yet there is no evidence to suggest that any of these policies had any effects on the level of the state's fishing activity, generally, or the shrimp industry, specifically. Annual shrimp landings continued to increase and, even as early as 1918, began to approach one million pounds (Chestnut and Davis, 1975:132).

The period between the world wars was distinguished by some spectacular leaps in the development of the fishery in North Carolina. But even more, it was an era of exploration as fishermen, dealers, and state officials experimented with capture techniques, marketing strategies, and legislation. We will examine this throughout the remainder of the book. Both successes and failures occurred, and the uncertainties and vicissitudes of the economy during the Great Depression and its aftermath framed many of them. Yet enough successes were seen to create the underpinnings of, and then sustain, the economic and cultural framework of a very durable and important fishery.

EARLY MORNING RUDE AWAKENINGS During the post-World War I years, the shrimp industry continued to expand rapidly in the United States. By 1931 it was sixth in volume and seventh in value among the food fisheries. Although North Carolina's contribution remained low during these years (less than one percent), the state's version of the industry began to prosper. The peak year was 1923, when more than 1.6 million pounds, valued at

$51,000, were captured. In that year 998 refrigerator cars were used to carry shrimp to markets in the Northeast. By 1925 over three hundred fishermen, nearly all from Brunswick County were targeting shrimp.

As we indicated previously North Carolina also eventually benefited from the use of refrigerator cars. Although the concept of freezing fish for transport to market was introduced in the nineteenth century, its use did not spread into the South until the early twentieth. And even as late as 1926 only little more than one percent of all frozen fish in the nation came from the South Atlantic and Gulf states. In that year, of the 169 packinghouses handling frozen fish in the country, none was located in North Carolina (Report of the Commissioner of Fisheries, 1926; and Lemon, 1936). Refrigeration in North Carolina during the period was used primarily to produce ice for shipment of fishery products. This practice continued through the 1960s. Trucks were introduced as the primary mode of shipment to replace refrigerated railway cars in the mid-thirties. At that time, twenty hours were required to transport fishery products to New York City.

Even though the North Carolina shrimp fishery was undergoing steady expansion, because of the percentage of catch relative to the total national production (about one-half of one percent) and that of the South Atlantic region (about 1.5 percent) in 1931, Johnson and Lindner were prompted to proclaim that the Tar Heel state was of "no great significance" in the nation's industry. Eventually, North Carolina's contribution to the South Atlantic regional catch (North and South Carolina, Georgia and the east coast of Florida) fluctuated greatly, but reached an average of 24%.

During the twenties and early thirties, Brunswick County continued to be the center of the North Carolina shrimping industry. In Southport alone, sixty-two boats, along with those coming from other areas, were in the harvesting sector in 1932. This generated employment for somewhere between five and six hundred people, including more than two hundred seasonal and part-time workers in the packinghouses (*Morning Star*, August 15, 1921, and April 18, 1932). The number of packinghouses increased to nine and the number of boats to two hundred during the peak season in the mid-thirties. Shrimping had become the community's most important industry and began to dominate its way of life, according to people I interviewed in the early 1980s who had been part of the growth of the fishery during that exciting period. The small port's population mushroomed during the season with hundreds of migrant fishermen arriving, many of them bringing their entire families.

The editor of the Southport *State Port Pilot* recalled his first four months there in the mid-thirties. Living on the waterfront, his sleep was "rudely shattered by the discordant sound of many noisy gasoline engines" as the trawlers moored along the docks in the harbor prepared to leave in the early morning.

Shrimping was still conducted exclusively in the daylight hours. Ice would be loaded on boats either from the packinghouses or trucks before they left the dock. The boats usually returned to port early in the afternoon because of the perishable nature of the harvest. By nightfall, all the catches were in, and at the nine packinghouses, several hundred "pickers," most of whom were African-American women, began to head the shrimp. Several of the creatures could be held in each hand simultaneously and the heads flicked off by the ends of the women's thumbs. Their hands would rise up from the holding tubs and, with a series of lightning motions, the heads would be popped off. While their hands were still moving, never stopping, they would direct the headed shrimp toward the washing vats. They would drop the headed shrimp, and again without stopping, move toward the holding tubs, all in a machine-like circular motion.

Local residents brought to our attention, and newspapers verified their accounts, the fact that "heading" would continue late at night, frequently into the early morning hours. There was much singing, referred to in North Carolina as "chanties," with "real melody floating up from the picking houses" (Southport *State Port Pilot*, July 27, 1938). The "pickers" were paid *a nickel a bucket*, and it is to honor those wonderful pioneers that this book was titled after their efforts. Life was not bad, if not entirely good! When the shrimping season began, the *Pilot* (September 28, 1938) proclaimed:

> It is no news that all of the shrimping packing houses pay off their labor each day with nickels. It is all piece work, the completed piece being a bucket full of headed shrimp and the pickers receiving the nickel as fast as the buckets are delivered at the washing vats. To carry on the business the banks will soon be calling for thousands of nickels daily from other banks. The most common form of currency for barter at the stores will be nickels. Some of the nickels will pass and repass through the wet hands of the pickers ....

After the shrimp were headed, they were packed in boxes with ice and stored in trucks. When the trucks were fully loaded, frequently early in the morning hours, they left for the markets.

Boxes of sorted shrimp ready for heading, packing and shipping.

CARTERET COUNTY AWAKENS Brunswick County's near monopoly of the state shrimping industry began to erode in the late 1920s as interest in the fishery began to spread to northward to Carteret County in the Morehead City, Beaufort, and "down-east" coastal littorals (Carteret

County *News Times*, October 11, 1955). The first activity in the shrimp fishery was an alternative to slack times in other fisheries and farming. A Florida firm located a packinghouse and six trawlers at Beaufort in 1933 (*Atlantic Fisherman*, November 1933). The newly appreciated abundance of the shrimp stocks and closer access to markets combined to create conditions conducive to rapid development, once it got underway.

Heading Shrimp for a Nickel a Bucket.

At first, the fishermen in Carteret County joined other migrants *in* Southport to master the skills of shrimp harvesting. When they returned home, they carried nets and otter trawl doors with them. At about the same time, an enterprising firm from Baltimore, which could see the promise in central North Carolina's *inside* waters, sent an experienced shrimp-boat captain from Louisiana to demonstrate the craft to fishermen in the nearby Pamlico Sound area. Under his guidance, nets and doors were obtained from Louisiana, and a number of local boats that had been used primarily for pulling long-haul nets and oystering in the sound, were rigged for shrimping (Carteret County *News Times*, October 11, 1955).

It was during this period that shrimping became firmly entrenched in the *annual rounds* of cash-producing activities in the Pamlico Sound and Carteret County areas, as it had earlier in Southport. Fishing and other activities became organized around the shrimping season. This included work in non-fishing jobs, later including government work for those employed at the Cherry Point Marine Corps Air Station, the Division of Marine Fisheries in Morehead, and the NMFS station in Beaufort. Vacation time, sick leave, and personal-leave days were scheduled to take advantage of the peak abundance periods for those who had grown up in fishing families and saw fishing activity as an important supplement to their incomes, as well as an important feature of their culture. For the full-time commercial fishermen, boat building or repairs, farming, home repairs, and even community political activity began to revolve around the increasingly lucrative shrimp harvest, processing, and marketing.

Wives and children became "headers" (the term "picker" was used more frequently in the southern part of the state). Wives also served as bookkeepers for their fishermen husbands and, eventually, as representatives to important public meetings on resource-management issues and regulations. According to one of our respondents in Pamlico County, these patterns became prevalent among some fishermen who, at first, did not eat shrimp "and laughed at people who did." Obviously, this attitude changed as the economic returns from shrimp harvesting began to improve the lives of the fishing families.

The director of the state's Division of Commercial Fisheries wrote in his 1940 report that as many as a thousand people had directed their efforts toward harvesting, processing, and marketing shrimp. He noted that while Southport had been the early leader in shrimping activity, "for the past several seasons it has been gaining in Pamlico Sound and Cape Lookout." Earlier, the Raleigh *News and Observer* (November 12, 1939) noted, "A million pounds of shrimp were taken during the past week from coastal waters between Albemarle Sound and the South Carolina border.

More than half of the total poundage was taken from Core and Pamlico and adjacent waters."

As shrimping gained popularity in Carteret County during the thirties, a unique method of harvesting developed near the small village of Harkers Island. Its development has had a profound influence on one portion of the coastal region where it is pursued even today. The introduction of "channel-netting" in many ways reflects the dynamic social and technological forces that have shaped North Carolina fisheries since the advent of commercial shrimping. Because of this, a detailed account of channel-netting is presented in the next chapter.

Interviews I conducted with people who participated in the early development of the fishery produced convincing evidence that shrimping became popular with the fishermen in the mid-coast counties because the season came between the spring oystering and the late-fall fin-fishing seasons. Smaller boats could be converted for shrimping and then re-rigged for the other harvesting activities.

Switching activities, which began to emerge as early as the 1920s, now became firmly established in the 1930s. Shrimping added a new dimension and added economic opportunities during the Depression years. Because of the migratory nature of shrimp, work patterns developed so that fishermen could either follow the shrimp as they moved geographically or change from one type of fishing activity to another.

The notion of switching activities or changing geographic locations is referred to as "annual rounds" and has become a permanent fixture in the North Carolina maritime landscape. However, changes do occur *within* the annual rounds for some types of fishermen as opportunities develop, which I will explore in a later chapter. In fact, some state fishery managers have referred to North Carolina fishermen as "opportunistic" in a complimentary sense. And, as we saw in the first chapter and shall see again later, the willingness of North Carolinians to adapt to emerging opportunities has angered fishermen from other regions who, under the guise of resource protection, have put a great deal of pressure on regional managers to eliminate competition.

As the fishery developed in the thirties, shrimpers from Florida, South Carolina, Georgia, and North Carolina concentrated in Southport in late summer and then moved farther southward for the winter harvest. In a 1938 presentation to the U.S. Bureau of Fisheries, Milton Lindner, in charge of a shrimp research project for the U.S. Bureau of Fisheries, took note of migration of fishermen to Florida as the concentration of shrimp moved to the Sunshine State. After first moving into South Carolina and Georgia in late summer and early fall, the fleet was found to have migrated to an area between St. Augustine and Cape Canaveral.

During this period, the majority of North Carolina fishermen, however, fished for *only* shrimp during the height of the season, and this practice continued until the nineties because it was so lucrative in comparison to other fishing activities. Until recently, those who did not migrate elsewhere normally engaged in other fishing activities like sink-net fishing for mullet and croaker. This has changed somewhat to include crabbing, as we shall see in later chapters. Early in the nineteenth century, sink or "drop-net" fishing had become an extremely important activity in Carteret County during late fall and the winter months. At first it was for subsistence or local marketing. With the development of the "mullet line," extending rail transportation from Morehead City to the state's interior at the end of the century, fall fishing became an integral part of the annual rounds of the central-coast fishermen (Maiolo and Tschetter, 1982).

In the Southport area during the thirties, another interesting pattern developed for some of the fleet. Depending upon the season and the abundance of shrimp or fish, some of the mid-sized vessels were used for both shrimping and sportfishing. For these boats, the annual round took on the unusual characteristic of switching from commercial to sportfishing, and was later adopted in other areas (e.g., Manteo).

Chapter Two

IMPROVED CAPTURE TECHNOLOGY The growing popularity of commercial shrimping was associated with improved technology in vessels and gear. Vessel types continued to range from decked craft to open skiffs. In 1936, the Southport *State Port Pilot* noted that Carteret County boats were of the "blue fish type. They are adapted for any type of fishing and need only a change of rigging to go from one thing to another" (September 23). However, two basic types of shrimp trawlers began to dominate the 1930s scene. With most of the shrimping taking place offshore in Brunswick County, even though the trawling radius was limited to a few miles offshore, Florida trawlers became quite popular. This design has the engine room forward and the fish hold in the stern. It was specifically developed as a shrimp trawler and is still the predominant large vessel in the industry. Boats in this class ranged from thirty to forty feet during the period under discussion.

In Carteret County, the "Core Sounder," so named because it is generally believed to have originated in Core Sound communities, became quite popular as a shrimp boat. With its flared bow and rounded stern, it was (and is) considered an excellent choice for trawling in shallow estuarine waters. They range in size from thirty to forty feet and are powered by gasoline engines. Most of them were built in the communities of Harkers Island and Mashallberg. Some Florida type vessels were built there also, as well as in Brunswick County.

Nearly all the work on boats, including hauling in the trawl, was done by hand until the late thirties, when some of the larger vessels were equipped with power winches. They are operated either with a separate deck engine or a power takeoff from the main engine and are used in conjunction with a mast and boom or a mast and "A" frame with rope towlines. The separate deck engine and "A" frame were gradually replaced by winches powered by the main engine and used in conjunction with a boom.

The technology of locating the shrimp also evolved over the decades. When trawling was first introduced, it was a hit-or-miss operation. A number of boats would go to a particular location and search until shrimp were located. In the Brunswick County area, traditional long-haul seines were used on occasion to determine if a location was promising. Further north, pound nets were used for this purpose.

Carteret County fishermen followed a more primitive method of dragging a long oar through the water while the boat was running at slowly. The presence of shrimp could be detected when the moving oar disturbed them and caused them to jump out of the water (Carteret County *News Times*, October 11, 1955; interviews with fishermen). Small "try nets" were first introduced in the Gulf of Mexico and Florida waters, and by the late-thirties, they were beginning to be used in Southport.

Shrimp nets were originally made in either Florida or Louisiana. In 1934, Lewis Harden, a native of Florida, arrived in Southport, opened a packinghouse, and showed local fishermen how to make their own nets. Until the 1940s, however, fishermen farther north continued to purchase their nets from elsewhere (Carteret County *News Times*, October 11, 1955).

The growing popularity of shrimping and the prospects for good earnings were not without problems. As the fishery began to mature in capture technology and marketing, the capacity of the developing social and economic structures to deal with such problems lagged behind. We will explore these matters following a brief visit to the most intriguing culture of channel-netting in the next chapter.

Laying on a Set: The Art of Channel-netting[1]

Throughout my fieldwork experience in virtually every fishery in the southeast U.S. since 1977, I have always been impressed by the remarkable capacity North Carolina fishermen have for innovation. One particular method of catching shrimp, called channel-netting, is a good example. It is also a good illustration of how customs governing fishing develop in a way to accommodate the needs surrounding production activities with folk concepts about the ecology of the resource.

Historically, channel-netting has been a local phenomenon pretty much restricted to the villages of east of Morehead City in Carteret County. Its development is one of the reasons for the county's entry into the shrimp fishery in the first place.[2] I have selected channel-netting as a special case study to highlight the technological and cultural dynamics of fishing as an industry and a way of life.

Channel-netting is a capture method that never contributed more than a small amount to the state's annual harvest. Still, it is one of the few methods that can be characterized exclusively as North Carolinian in both its invention and use.[3] And, until recently, it has been an extremely important style of fishing and source of income for fishermen in the villages "down-east" in Carteret County. Until the mid-1980s, about one hundred families depended heavily on income derived from this technique. While a comparatively small number of shrimpers in the entire state have relied on this harvest method, in the villages extending from Harkers Island to Atlantic as many as 20% have done so.

First used in the 1930s near Harkers Island, the channel net is stationary in use, unlike the otter trawl, which is pulled behind a boat. Contrary to the findings reported in another writer's study (McKenzie, 1974:50) the channel net was found to be very productive for all three species during the entire season. Though the technology has its closest affinity to the otter trawl and shares some of its characteristics, there are earlier predecessors in design and method of development. The earliest was the "drag net" (or crabshore net), a seine fished by two men from a single boat. This was first developed and used in Carteret County in the early 1800s. Ranging from eighty to one hundred yards in length with staffs placed on either end, the drag net initially was used in the crab fishery and near the shoreline. The folk theory of the technique was that crabs would be carried along the tidal flow as they migrated seaward and would be easily caught by placing the net in their paths. Then the outermost portion of the net would be brought ashore "upcurrent" of the other end, the entire net was brought ashore, the catch emptied, placed in a small skiff, and the process repeated until either there were no more crabs or the fishermen's day ended. What is important about this technique is the use of tidal currents to essentially do the fishing, which carried over to the channel net.

With the expanding shrimp markets in the 1930s we discussed in the second chapter, the drag net was adapted and used in a similar fashion to its use in the crab fishery. Further improved and re-labeled the "shore net," it was set from the water line out to 120 yards. As otter trawls became more widely used in the mid to late 1930s in the Harkers Island area, fishermen there were very aware of the strong tidal currents near Shackelford and Core Banks, especially on the ebb tide. Accounts differ about the exact time, but interviews Marcus Hepburn and I had with knowledgeable fishermen in the 1980s indicated that two fishermen decided to use the otter trawl in a stationary position. Two skiffs were anchored crosswise to the tidal current and then the trawl was stretched between them. The net was fished by periodically pulling up the tail bag (bunt end) and emptying it. This way of using the trawl net did not last too long and was discontinued in favor of using an adapted version called the "pocket net," a piece of small-mesh netting fifty or more yards long. At the center, a diamond shaped hole was cut and a simple four-sided pocket (bag) was attached. The net was fished by pulling up only the end of the pocket and dumping the catch into a skiff placed *over* the pocket. This meant the entire net did not have to come up in order to harvest.

Net technology underwent more modifications, evolving into what is now known as the channel net. The modern version is set on the stern end of the boat, with ends of the staffs running across the boat and resting on the gunwales. As the ebb tide begins to "set up" or run, the fisherman first locates the edge of the channel or bank. This is usually accomplished by using a long pole to test depth. The edge of the bank is marked with a buoy. Then, making allowances for anchor line and bridle length, the fisherman throws one of the anchors overboard and sets it firmly. The net is then pulled off the stern so that approximately one-third of it will extend off the bank and into the channel. This is the maximum permitted by regulation. In most instances, the net is deployed from the back (shallows) and into the channel. Today the net can measure no more than forty yards from one wing to another so it does not interfere with navigation.

The channel net is a fairly simple and straightforward gear. But its successful use is based on a comprehensive knowledge of the gear, the characteristics of the bottom (where, for example, the channels or currents run), a large reservoir of information about the behavior of the target species, and the demonstrated understanding and compliance with the customs that govern the behavior of the fishermen. To actually fish the net, the fisherman stations himself (there are no women channel-netters) to the leeward side of the buoy, marking the line to the extension, with that line tied to his boat. Depending on the circumstances, the net is

Laying on a Set: The Art of Channel-netting

*Sketches to the left illustrate channel-net technology;
the one to the right illustrates laying on a set.*

fished in fifteen- to thirty-minute intervals. It is hard and lonely work. Especially when there is little wind in the middle of the summer, flying insects of several sorts can make a night less than pleasant.

And it is backbreaking. When Marcus and I went with and observed a "highliner" in action (someone considered to be one of the best), we hit a run of shrimp of major proportions, requiring the net to be pulled about every five minutes. That night, the boat landed more than fifteen hundred pounds of shrimp and at least five hundred pounds each of finfish and crabs. So I can verify from first-hand experience that this is very hard work. We finished at 3:00 A.M. so the fisherman could then transport his catch to the dealer. I can only imagine what that night would have been like had our fisherman been working alone. But somehow they manage.

The line attached to the extension is pulled when it is time to take a look. When the webbing is clear of the water, the entire front of the extension is pulled up. The shrimper then works back toward the tailbag by shaking the net so that shrimp or other fish and the debris will be moved aft. When he reaches the bag, he brings it on board and dumps the catch on a culling platform set up at the stern. The culling platform reaches across the entire stern about a foot lower than the gunwales and extends forward to the first thwart. The net is returned to the water and the fisherman then culls and packs his catch while he waits to take another look.

Channel nets are usually fished by one person. A family member or friend may tag along occasionally for a night's visit. In the past, a son may have accompanied his father to learn his future occupation. Marcus and I discovered several cases in which two fishermen worked the same net as "partners," or two men may decide to be "mates." In the latter case, separate nets are fished by each of the fishermen, but the catches are combined and sold together, with proceeds divided equally. The mates share information, take turns loading the catch, and generally keep a watch on each other. They usually fish in the same general area.

WHERE, HOW, AND WITH WHOM This fishery begins in late spring and continues through early fall. Some variations on this schedule occurred during the early history of the fishery. In those years, channel-netting was conducted only during the spring because the fishermen pursued other fish during the late summer months. By the mid to late fifties, it was discovered that the gear was productive well into September. When the fishing productivity drops off temporarily in August because there are fewer productive locations to "set," some net fishermen take the month off to attend to other duties and then return in September.

The fishing occurs in a limited number of locations. In the down-east region, local customs, even today, allow these locations to be essentially "reserved" for the fishermen from nearby villages. Fishermen from other villages (farther west) can fish for shrimp in the area, provided they do not channel net and do not interfere with the local channel-netters. These are local customs, mind you. They have no basis in law, but are respected, nevertheless. I will return to this topic at the end of the chapter.

As the evening's activities wear on and as other fishermen bring their boats into the areas (always shrimp trawlers), the channel netters turn on a spotlight and move the light in a direction around them that they expect the trawler to follow. Everyone in the industry seems to understand these and other customs, including local law-enforcement personnel, and no one interferes. Breaking these and other local rules outlined in the next few paragraphs can result in anything from ridicule to physical threats.

Channel-netting requires, in addition to an abundance of migrating shrimp, a relatively strong tidal current on the falling tide and a bottom free of debris. In places where exceptionally strong currents occur, fishing is either avoided or a smaller net is used. Historically, the most successful locations around Harkers Island are found in waters called the "Straits," a virtual "river" between the island and the mainland and connected to Core and Back Sounds. There are four locations in the Straits having a recognized and named status (such as the obviously named "Gold Mine"). Several other locations near there and to the west near Beaufort and Morehead City are used intermittently, as well, but are not named. Most of the set locations in Core and Back Sounds do have names. Most of them are named after the person who first started setting there or who was otherwise traditionally associated with that set. Some locations are known to be better producers than others, and, as a consequence, a complex system of local rules governing use of the "good sets" has evolved.

Local fishermen who wish to fish at the well-known and productive locations must "lay on a set." To do so—that is, to reserve it for use—the fisherman must have a skiff, with his channel net gear on board, at the location. In addition, traditionally someone has had to occupy the boat, though not necessarily the fisherman. In some cases, the fisherman would hire a young boy to perform that task. The boy was taken to the fishing boat in early morning and remained there during the entire day. The fisherman would return before dusk to take over. Someone would usually remain in the boat even while the young man was being ferried to shore.

If a fisherman wished to work a particular set but someone else had already begun fishing or laying on that set, he could, in effect, get in line by placing his skiff and gear behind the other boat(s). If that occurred, then the first boat in line had the right to fish the set that night but had to relinquish the set for the person behind him the next night. I have seen as many as four boats laying on a set in some of the prime locations. But the important thing to remember is that the person waiting has to be present when the other fisherman takes his net out of the water after the night's work. Even though a fisherman might have been waiting in line all day and all night behind another who was fishing, if he leaves, even for a short while, when the person fishing takes up his net, then the latter has the right to retain the set for the following evening simply by remaining on location.

In the late 1970s, the rule governing presence of a person in the boat while laying on a set was challenged. Some argued that the owner, or a paid helper, should no longer be required to in the boat to be considered as laying on a set. Catches had been poor, and the cost of hiring young boys was rising. If all of the other rules were followed, it was suggested, the first boat in line would have the right to fish the first night, followed by the second, third, etc. But most fishermen objected to a change, arguing that the stricter rules, which required the investment of time and capital for the most productive sets, would keep out all but the more traditional channel-netters. Relaxation of the rule would permit casual fishermen to reserve a set without any investment of time, energy, or money, other than putting his skiff in line and returning when it was his turn. The proposed changes were not adopted.

In January 2003, just as plans were being laid for the upcoming spring fishing, the "person-in-the-boat" rule was still in force, but support was weakening. In spite of 2002 having proved to be a very good year for shrimp harvesting, as we shall see in Chapter Six, the channel-netters did very poorly. Even the most historically productive sets did not yield good results. Most of their catches were small shrimp, bringing about one dollar a pound at the dock, about a fifty-year low.

As I mentioned earlier, another important custom has to do with the distance between fishermen at work. It is considered inappropriate for someone to trawl nearby or set another channel net within a quarter of a mile. One night, Marcus and I observed some trawlers from another part of the county, the village of Salter Path on Bogue Banks (considered to be interlopers even under the best of circumstances), working closer than a quarter of a mile from channel netters. We were in my boat maneuvering around and observing the harvesting. When the shouting started, we tried to put some distance between us and the potential combatants, only to find our escape route cut off by another channel-netter. So our retreat required coming within the coveted quarter of a mile. As we passed by, the fisherman yelled, "It's a God-damned good thing I know you two dingbatters (a pejorative term for outsiders), or I'd be blowin' your damned heads off." We took this comment as a guarded jest, since we knew the fisherman, and he knew us. Marcus and I both had small homes on the island, his permanent, mine of the vacation type. Both of us were professors at East Carolina University and were well known to the islanders as kooky, maybe, but trustworthy, nevertheless. Marcus, an anthropologist in traditional form, chose to live in the field where he did his work.

The less-productive locations have had fewer rules, and these do change from time to time. The netters who take part in rules discussions are only those who will be fishing the locations where the rules apply. Also, over the years, some traditional locations have been identified with kinsmen or friendship groups who seldom fish in other locations. Likewise, others do not fish in those traditional areas.

During the recent past, a substantial channel-net fishery has developed in and near the New River at Snead's Ferry in Onslow County, next to Camp Lejeune. Somewhere between fifty and sixty vessels operate between there and Swansboro to the north; and Topsail Beach to the south. Management of the shrimping in this region is directed mainly toward separating different user groups by space and, to a degree, times when fishing is permitted. In contrast to down-east Carteret County, the channel-netters do not have a queuing system but do establish territoriality in areas where they are allowed to set their nets without blocking trawling or normal movement on the Intracoastal Waterway (ICW). Channel nets must be fifty feet from the ICW centerline, which forces them to set near the edge of the channel.

This means that some portion of the net will be in water too shallow for trawling. They do this by either installing screw-down anchors or by dropping an automobile engine block to the bottom to serve as an anchor. Then they attach identifying buoy markers to them.

Channel-netters have the advantage of fishing the entire week by virtue of regulations that restrict trawling, but not channel-netting, on the weekends. Thus, they can fish when the trawlers cannot, prompting complaints from trawler captains that the channel-netters have an unfair competitive advantage. The channel-netters counter that there are only so many nights per month when the tide is right for them to fish.

THE ECONOMICS OF THE FISHERY Among the more attractive aspects of channel-netting is its comparatively low cost of operation (i.e., expenses for gear, gasoline, and boat size). Since the net is not towed, there is no appreciable fuel bill other than getting to the fishing site, nor is there much wear and tear on the boat. There are fewer gear failures and snagging incidents, which can destroy a net in seconds. There is less noise, absent the constant drone of engines from other vessels in close proximity for hours on end during the tow process. And most channel-netters dock their boats within a few miles of their net locations. The boats are used only for getting to and from the fishing locations and taking the catch to market.

A new net—excluding lines, anchors, staffs, etc.—still can be put together for a few hundred dollars. A lot of the gear is fabricated by the fisherman themselves to save money. My fieldwork in 2003 did not produce a single new, or even close to new, boat in the channel-net fishery.

Catch totals for the whole channel-netting culture seldom reach four hundred thousand pounds. The best catches usually come in June. Marcus has seen one crew produce nearly seven thousand pounds over several days. Obviously those kinds of numbers are quite unusual. The catches at the better sets will be a hundred pounds, not more than five or six hundred, during the height of the season. At other times and places, a great catch is in the fifty-pound category, with less than fifty being more likely. When things are really, really good, a steady yield of six hundred to a thousand dollars per week, after expenses, would be considered to be quite acceptable, even in today's dollars.

This fishery is in decline, unfortunately, in Carteret County. The fishermen blame the poorer catches on the fact that large and efficient steel-hulled trawlers capture the best of the lot in the Pamlico Sound before they get to the channel-netting waters. Discussions I have had with a scientist at the DMF suggest that this kind of preemptive harvesting may be possible, but this could not be confirmed by empirical data at that time. He indicated that effort in the Pamlico Sound has not changed that much to have caused such a drastic decline in channel-net production in the Harkers Island region.

The down-east channel-netters complain, too, that even when catches are fairly good, ex-vessel prices are depressed. I have confirmed this through my own observations. It appears that imports of shrimp from other countries play an important role in this. I will deal with this matter in the sixth chapter.

There are now only between thirty and fifty channel-netters left on Harkers Island, and many of them are in the fishery on an opportunistic basis; that is, they fish only when they know things will be good or when the alternatives are even worse. My favorite highliner left the fishery in the late eighties. And while he still fishes commercially, a great deal of his effort goes into a small business he has created. Some fishermen have retired and have not been replaced in the fishery. Others left because they felt they could no longer support their families and either tried other types of fishing or got a "land" job.

There are just a few young men in the fishery on Harkers Island, which is indicative that labor recruitment into the system is poor. The young men from the down-east area who can afford to remain there allocate their efforts to more productive types of fishing. The old fish buildings and marina "hangouts" are gone. Most of the current netters go to a new, large, spacious, and well-stocked variety store for their morning coffee, sausage dogs, and conversation. It is there, too, where they stock up on their provisions for the next evening's fishing—packaged cakes and soft drinks.

But, in addition to the changing economics of the situation, the down-east fishery is in decline for other reasons. For example, the locally determined rules are receiving a lot of attention. Fishermen have left the group because they rejected the complexity of rules associated with laying on a set. Some have abandoned channel-netting because of some near-disastrous confrontations over the rules. There is no doubt in my mind that the economics of the fishery, namely ex-vessel prices, have played a key role here.

In regard to the role and function of locally derived controls, my friend Jim Acheson argued in 1975 that these kinds of customs are beneficial to both the fishermen, for obvious reasons, and to the management framework. Local controls often do have a resource-preservation component and are more acceptable to local fishermen because the rules are their own. He sees this as kind of a complement, or even as an alternative, to management schemes imposed by government agencies. At a recent meeting of the Southern Division of the American Fisheries Society (February 16, 2003), Mark Imperial and Tracy Yandle suggested some additional advantages of locally derived management over other approaches: e.g. the preservation of local cultural traditions, values, and small-scale operations; minimized environmental impacts; a valuable correspondence between rules and local conditions; and even the probability of fewer incidents of cheating. But they also pointed out that locally derived rules have the *disadvantages* of concentrating power in the hands of a few without external accountability and rules that are sometimes unsafe and that do not adjust well to changes in technology, stock fluctuations, and outside threats. Their discussion seems to have a great deal of relevance to the situation in the down-east region insofar as it is clear to me that the decline in the fishery is, indeed, partially accounted for by the disadvantages to local control that Imperial and Yandle enumerate (Imperial and Yandle, 2003:5).

Finally, in addition to the effects of real price declines I have already discussed, there is another external factor that bears heavily upon the down-east culture, generally, and that, in turn, specifically affects traditional commercial fishing activities such as channel-netting. I am referring to the change in cultural setting from one combining almost exclusively commercial fishing with some recreational fishing to one that is in the process of *reversing* those emphases. Building lots once valued at around $5,000 in the 1970s now sell well into the $60,000 range. And anything even remotely near the water *starts* at well over $100,000. Children of fishermen, if they choose to stay in the area, and many do not, have to find building lots and homes in distant villages on the mainland. In a few cases, they buy from their parents or other close relatives (and are financed by them) so they can remain on the island. This is such a force for change with so many implications that it needs to be dealt with in even more detail, and I will do so in the final chapter.

Thus, it appears there are a number of economic and sociological forces at work to put the Harkers Island channel-net fishery in jeopardy. Only time will tell if it can survive. Let us now return to a discussion of the development of the trawl fishery into a full-blown industry with all of the social, economic, organizational, regulatory, and political trappings that implies.

Shrimping Comes of Age in North Carolina

(WITH WILLIAM STILL¹)

Capture technology improved, more fishermen entered the industry, and harvest levels climbed through the 1920s. But then the onset of the Great Depression severely reduced profits. The thirties began with poor harvests, but after 1933, production steadily increased. The quantity went up dramatically in the state from 338,000 pounds in 1931 to over 2.5 million pounds in 1934 and then to nearly five million pounds in 1939. While only sixty-two boats shrimped from Southport in 1931, more than one hundred were in business by 1933. And three years later, there were more than five hundred shrimpers. The Wilmington *News*, in its September 27, 1934, issue, noted:

> ... with a big fleet of boats at work and more coming in daily, the catches of shrimp in the Southport area Monday and Tuesday were the largest in many years. Tuesday night the packing houses became swamped with the product and fearing too much of a congestion of northern markets, the boats were all ordered to remain in and not fish today.

Despite this "boom," discontent became widespread among Tar Heel fishermen throughout the decade. By 1938 Southport shrimpers went on strike, refusing to take the boats out to sea (reported in both the *Morning Star*, September 18 and October 26, 1938, and *State Port Pilot*, October

26, 1938). Low prices received from the dealers for fish and shellfish was a major reason for the discontent. All seafood combined brought an average of under *one cent* per pound in 1932, as compared to 1.9 cents in 1928. The average annual earnings for a North Carolina fisherman in 1931 were a paltry $169. And even though the prices and wages would gradually climb, the actual return to fishermen, in terms of real purchasing power, would be very little. It took the outbreak of World War II to spur improvement (for further discussion of this topic, see Maiolo and Still, 1981; and Still, 1987).

We found that each party in the industry blamed the others. Fishermen who used long-haul seines were criticized for glutting the market with product. Dealers and packers were blamed for encouraging the large harvests in order to gain a competitive advantage. And dealers blamed the government for *not* taking action. One dealer even blamed the situation on truck peddlers who sold the seafood inland. In February 1931, a Mr. M.S. Lee wrote to Mr. Luther Hamilton, "Why it is absurd to take seafood in an open truck and peddle it all over the country in the sun and rain …. Such is not sanitary," he complained (taken from the papers of Governor Max Gardner, North Carolina State Archives, Box 80).

THE DEVELOPMENT OF COOPERATIVES Inevitably the fishermen turned to the government for assistance. Early in October 1934, according to the records of the North Carolina Department of Conservation and Development, a committee of fishermen from the coastal counties met with the governor and urged "special action" to improve their plight. State officials conferred with representatives of the U.S. Bureau of Fisheries and governmental relief agencies. In December a plan was presented to the fishermen to organize a cooperative, which turned out to be a colossal misadventure. The idea was based on a similar organization that had been formed in Maine with the help of the federal government. Under the proposed plan, the self-help Division of the Federal Emergency Relief Administration (FERA) and local fishing communities would cooperate to establish facilities for handling, processing, and marketing seafood products. These facilities, which would be purchased with a loan from FERA, would become the property of the fishermen who agreed to join the cooperative (Department of Conservation and Development Archival Records; articles in the Raleigh *News and Observer*, January 1 and 3, 1935).

During the early months of 1935 the state's fishing communities were surveyed and over fifteen hundred fishermen agreed to join the cooperative. The organizers attempted to gain the support of dealers by emphasizing that the coop would not compete with them and do business with traditional markets in the northeast, but would develop new markets in the interior of the state. Nevertheless, a large percentage of the dealers opposed the cooperative concept, in spite of (and maybe because of) the support it received from state officials. H.F. Prytherch, Director of the U.S. Fish Biology Lab in Beaufort, was supportive, as well. He suggested that it would increase employment in the industry, stabilize prices, and, in a letter to a Mr. John Sikes of FERA (February 8, 1944, North Carolina State Archives, Box 44, RG69), emphasized it would "ensure equitable returns to fishermen. Further, a "very commendable feature of your program is that it will not offer serious or unfair competition to private concerns ... in the fishing business, but will ... benefit them by providing facilities for preservation and storage ...." The state also agreed to purchase substantial amounts of seafood from the coop for its educational and correctional institutions (letter from A. Brower, Division of Purchases and Contracts, to Mr. Etheridge, North Carolina State Archives, Box 44, RG 69).

Organizers of the coop applied for a loan of $139,000 from FERA to purchase equipment and facilities, which was approved in July 1935. In recommending approval, one official noted, "North Carolina has no plant for the processing of [frozen] seafood products, and therefore cannot supply its inland market." He noted that this situation had resulted in restrictions of fishing activities and in the "utter waste" of one third of all seafood caught. Further, the "present per capita seafood consumption in

North Carolina is about one half of the national average." He also made note of the fact that many of North Carolina's fishing families were on relief, and there was little chance for them to gain subsistence through farming or other work. Of the more than fifteen hundred fishermen who had already joined the coop, "824 are now on relief, 519 are certified as eligible ... and most of the rest are on the borderline" (Jacob Baker, June 20, 1935, North Carolina State Archives, RG69). According to data we have been able to retrieve from state records, there were approximately five thousand people, including part-timers, engaged in commercial fishing in 1934.

With the support of local communities and the state, the newly created self-help coop began establishing its facilities. Morehead City (central coast), Manteo (northern), Southport (southern), and Belhaven (inland) all supplied sites, some funds, and tax incentives. The main plant was built in Morehead. It included a receiving facility, a cold-storage and freezing plant, canning capability, and offices.

A small fish-handling facility was established in Manteo, an oyster-shucking plant in Belhaven, and a canning plant for shrimp in Southport. In addition, the cooperative, which took the name North Carolina Fisheries, Inc., purchased a fleet of refrigerator trucks to carry the processed seafood to the interior markets. While the physical facilities were being constructed, the coop was being organized. Of the more than fifteen hundred applicants, four hundred were initially approved for membership. We could not determine what the criteria for initial membership were, but assume it was based on dependency on commercial fishing for a livelihood.

The Morehead plant went into operation early in October 1935 and was followed within a few weeks by the other plants. The coop operated in the red from the beginning. Losses exceed ten thousand dollars during the first three months. Some of this was the result of the poor 1935–36 harvest season. Although the amount of seafood produced by the

Morehead City plant would gradually increase because of the deficiencies of the other smaller plants, it would continue to lose money. In 1936, about a year after opening, both the Manteo and Belhaven plants were closed, and the closing of the Southport facility followed in 1938.

In the Southport facility, both fresh and canned shrimp were supposed to be processed. The cannery was never put into operation, and the processing of fresh shrimp was not profitable. North Carolina Fisheries, Inc., was unsuccessful in its efforts to increase the consumption of shrimp in the state. As a result, contrary to the originally stated intentions, the coop began to compete with the independent dealers for the northern markets, according to news accounts at the time.

As with any government program, the coop was controversial from the beginning. Most fishermen liked the idea because they would be paid a set price, usually slightly higher than the return from independent dealers. And fishermen who did not belong to the coop could still sell to it. The strongest opposition came from the independent packers and dealers, who claimed that the competition from the coop was unfair, especially when the coop began to sell products to northern markets after attempts to establish interior markets in the state failed. And, as I noted above, this put the organization in direct competition with traditional seafood dealers, packers, and processors.

The cooperative tried various methods to develop the interior market without success. Efforts included educational programs designed specifically for public schools, radio advertisements, and sending salesmen to call directly on retail outlets, such as grocery stores, hotels, and restaurants. The editor of the *State Pilot* wrote (December 1935), "The sales organization of the North Carolina Fisheries, Inc., [has] failed to provide the promised market within the state.... The reason that there was no demand for shrimp in North Carolina this fall is because 85% of the citizens of the state never saw a shrimp in their lives."

It is hard to believe, but in 1938, near the north central community of Weldon, North Carolina, a truck carrying a load of shrimp turned over and spilled its cargo along the road. Onlookers could not identify the cargo. And the Raleigh *News and Observer* (December 9, 1940) reported that "hardly more than an average of four or five hundred pounds of the local catch [of shrimp] is consumed in North Carolina." I think this may be an exaggeration on the low side, however.

With the failure of market expansion in the interior of the state, the principal markets continued to be Baltimore, New York, and Philadelphia during the thirties. The cooperative was closed down after the 1940 season, and, despite local efforts to revive it, it never reopened. The buildings were eventually sold. The freezing plant would reopen under private ownership in 1942, however. Although the coop never made money, the government insisted it had been a success. One FERA official suggested the goal of making the cooperative self-sustaining was less important than aiding some fifteen hundred families who had been on "relief." In response to an inquiry on the matter, President Roosevelt was informed that North Carolina Fisheries, Inc., was beginning to demonstrate a considerable amount of aid "to that section of the coast" (November 7, 1936, North Carolina State Archives, Box 44, RG69).

World War II ended the Depression and, for most Americans, there was a return to somewhat of an economic, if not political, normalcy. This may have contributed to the demise of the cooperative. The reason is that, as the increase in demand for food products developed nationally, fish dealers could demand higher prices in the marketplace. Without the competitive advantage of the price differential offered by the coop, it could no longer maintain the loyalty of the fishermen. Further, other government subsidies in the form of low-cost loans had an impact on the organization (*Southern Fisherman*, January 1943).

THE WAR YEARS World War II at first created severe problems for the U.S. fishing industry. The 1942 harvest level declined by nearly a billion and a half pounds over the previous year, and the source of the problem was not the condition of the stocks. Military needs resulted in a significant drop in the construction of fishing boats. In addition, some seven hundred vessels were taken over by the government for military purposes. Hundreds of others were laid-up at various times because of severe shortages in spare parts for engines and fishing gear. Fishing nets could not be obtained, and twine and cordage were scarce as well (*Southern Fishermen*, January 1944). Many Americans still do not realize that fishing outside of protective inlets had to be curtailed because of German submarine activity. In North Carolina, this threat extended along nearly the entire coastline!

Manpower also became a critical problem as thousands of fishermen were drawn into the armed forces and war-related industries. The federal government attempted to cope with this problem by deferring fishermen from the draft, but this met with limited success. In January 1943 North Carolina's Department of Conservation and Development recommended to the War Manpower Commission that Selective Service (i.e., draft) regulations be amended to allow the deferment of experienced commercial fishermen on the same basis as the nation's farmers. By August of that year the Selective Service System had placed some classes of commercial fishermen (boat captains) on the list of critical occupations. Three months later, however, the head of the U.S. Fisheries in Beaufort, North Carolina, was complaining that nothing had changed: commercial captains were still being drafted (letter from Mr. Robert Prytherch to Colonel Thomas Upton, Assistant Director of the State Division of Selective Service, November 2, 1943, North Carolina State Archives).

The developing labor shortage often resulted in the illegal employment of children. Shrimp packinghouses in Carteret County and Southport were charged with violating child labor laws by employing children under the age of sixteen. On the other side of the issue were the state and U.S. Fisheries Bureau officials who strongly supported the utilization of the underage children, pointing out that it was part-time employment, primarily seasonal, and that the manpower shortage was causing severe harvesting and processing problems that would affect the region, if not the nation. "In this county," Dr. Prytherch wrote to the Office of Fisheries, U.S. Department of the Interior, "there are approximately twenty concerns engaged in shrimp production and shipment, and there are no laborers to prepare this product if they are to be deprived of the part-time employment of children" (July 16, 1943). Child-labor restrictions were then temporarily eased, but the labor problem remained acute until hostilities ended.

Despite these problems, during the war years and the decade that followed, the shrimp industry began to emerge as the most important of all the state fisheries in terms of value. This was due to several factors: the growing popularity of shrimp with the consumer, marketing advances, improved technology, expansion of the areas where shrimp were harvested, and the relaxation of restrictions on seasonal and night shrimping.

One lifelong commercial fisherman I interviewed in the late seventies was certain that the remarkable increase in the popularity of shrimp was due to the influence of military personnel stationed in the state during the war. However, an important factor nationally was that, unlike meat, *seafood was not rationed* during the war. And, for the first time, there was a significant demand for shrimp in the interior of the state. Frozen shrimp were shipped as far as Greensboro and Charlotte, not just from coastal North Carolina, but from as far away as communities bordering the Gulf of Mexico. North Carolinians began to ship shrimp to Savannah and to Louisiana. In 1942, a Mr. R.R. Barbour purchased the freezing plant that had formerly belonged to North Carolina Fisheries and by the following year, was shipping shrimp "throughout the USA from Florida to Maine and the Middle West" (*Southern Fishermen*, January, 1943:14).

Although canning would continue on through the post-war years, the widespread adoption of quick-freezing shrimp was the most important marketing advance up to the 1940s. In the mid-thirties, North Carolina Fisheries had installed the first freezing plant on the coast. However, it utilized "sharp" or slow freezing, which was satisfactory for fish preserved whole, but not for those dressed of filleted, nor for shrimp. "Quick" freezing was developed primarily by a Mr. Clarence Birdseye, whose process was introduced by General Foods in the thirties. The name "Birdseye" became synonymous with quick-frozen food products. The general acceptance of frozen food by consumers occurred during the war and immediately afterward, when food shortages encouraged many (who otherwise might not have tried them) to turn to frozen food products.

Some fishermen, fish dealers, and experienced resource managers with whom we consulted during the seventies and eighties suggested that quick freezing was one of the most important milestones in the history of shrimping. The logic was that the creatures were no longer so perishable. Instead, shrimp became a relatively stable commodity, the price of which the producers could control by freezing and holding for the best price. Interestingly, this was the concept that the North Carolina Fisheries coop had envisioned in the thirties, but it took the war to make it a reality. Even though there had been an attempt in Southport to establish a quick-freeze plant in 1942, the war prevented the plan from being implemented, and it was not until 1946 that such a company was able to get underway in Beaufort, North Carolina (Carteret County *News Times*, October 11, 1955, and personal interviews with Mr. Clayton Fulcher).

An examination of historical events, however, indicates that a different story unfolded about quick freezing and the shrimp industry. It appears that the introduction of the new process did, at first, play a role in rapid expansion of the industry. But the bulk of shrimp continued to be shipped fresh to northeastern markets. And it appears that the advantage North Carolina had gained diminished as frozen shrimp from other sources became popular. William Ellison, Jr., warned, "North Carolina's producers and dealers are square up against the fact that they may have to freeze at home or suffer a severe setback in years to come" (Carteret County *News Times*, October 11, 1955, and Ellison's *Report on the Manufacturing Potential of the North Carolina Fishing Industry*, 1956, in Box 71, Governor Luther Hodges papers, North Carolina State Archives). In 1956, members of the Clayton Fulcher family built a quick-freeze plant in their hometown, Atlantic, North Carolina, and began shipping frozen shrimp in 1957. This, according to Mr. Fulcher, did not succeed for more than a few years. Afterward, and to the present, the Fulcher business became mainly a packinghouse (heading and packing shrimp and packing finfish, crabs, oysters, and clams). Other dealers changed over to freezing reluctantly. One of these, the Belhaven Fish and Oyster Company, switched back to fresh shrimp like the Fulcher Company.

Ellison's warning rang true throughout the remainder of the century and does so even today. Freezing capacity in North Carolina is still limited and puts the state at a tremendous competitive disadvantage. Innovations at the other end of the dock (processing and marketing) have not kept up with innovations in the harvesting sector.

Nevertheless, eating shrimp was growing in popularity among consumers, which led to a substantial increase in the fishery's market value. In turn, this led to more harvesting activity and landings. The industry experienced a rapid expansion within a very short period. After a poor harvesting season in 1941, in which the landings dropped to less than 2.5 million pounds, the harvest level stabilized in the range of 4 to 6 million pounds from 1942 to 1944. In 1945, the catch increased to 10 million pounds, one of the highest harvest levels in the state's history.

Shrimp production in the state peaked at more than 14 million pounds in 1953. That volume has not been achieved since then, with the years 1985 (11.6 million), 1991 (10.7 million), and 2000 (10.3 million) coming closest. By 1948 only menhaden outranked shrimp in value, but this would change within a few years, with shrimp becoming king. During this same period, the state's contribution to the South Atlantic region (excluding the Gulf of Mexico) steadily increased to the record setting year of 1953. North Carolina's harvest that year represented more than 36% of the region's catch. By 1952, shrimp had become the most valuable single fishery in the country, with over 225 million pounds harvested.

Chapter Four

The rapid increase in the demand for shrimp during the war years placed a heavy burden on the entire fishing industry. One dealer we interviewed recalled that "the shrimp were here, but we didn't have anything to catch them with." In 1944, U.S. war production had become so efficient that the government relaxed its control over the use of certain strategic materials such as lumber, tools, and machinery parts. This made them available to civilians on a priority basis, which boded well for the construction of new fishing vessels. At the same time, as the fear of submarine attacks subsided, there was an easing and eventual lifting of restrictions on fishing outside of the state's inlets. All of these factors came together to spur efforts to revive and rebuild the entire fishing industry in the Tar Heel state.

National Archival data and records from the North Carolina Department of Conservation and Development indicate that a tremendous shortage of fishing boats had developed. Several hundred boats were found to be about worn out. The director of the department was told that the federal government's Smaller War Plants Corporation would provide low-cost loans for the construction of new fishing boats. It estimated that loans could be approved for the construction of about eighty new boats per month. Local newspapers in late 1944 and early 1945 reported this. The *State Port Pilot* (July 1944 through August 1945) stressed the fact that many of the new craft were made more fuel-efficient by diesel power. At the outbreak of the war, there was only *one* shrimper using diesel power. Beginning in 1944, the number increased to well over half of the operating vessels in the eighties and has continued to increase to include a majority of shrimpers, according to interviews with fishermen. The larger new vessels that were built during the forties came mostly from Florida. But local builders prospered as well, especially in Marshallberg, Harkers Island, Southport, and even inland in Washington, North Carolina. There, a company named the Pamlico Shipyard received a lucrative contract from General Seafoods specifically to build shrimp trawlers. Between 1944 and 1946, that shipyard alone built forty-five forty-nine footers, three sixty-two footers, and one seventy-two-foot shrimper. Interestingly, however, most of those wound up being sold to fishermen in Louisiana.

POST WAR: IMPROVED CAPTURE TECHNOLOGY AND TERRITORIAL EXPANSION
The trend toward larger vessels was based on several factors. They can travel farther and stay at sea longer than smaller ones, they are safer, and they permit the fishermen to participate in many more fisheries up and down the coast, such as scalloping on Georges Banks, Massachusetts; finfish and shrimp trawling in the Carolinas; and calico-scallop fishing in Florida. Also, they are more efficient in terms of operating costs. A former manager of the Belhaven Fish and Oyster Company said, "You would have starved to death ... with the little rigs" that were used before the war.

The new diesel-powered trawlers were frequently equipped with much more sophisticated gear than the standard vessels of the thirties. Much of the new technology was adapted from wartime technological advances. Some surplus military vessels were purchased by fishermen in both Brunswick County (for trawling) and Carteret County (for menhaden fishing) that were equipped with radar, ship-to-shore communication systems, fathometers, range finders, and so on. The value of these kinds of equipment was quickly recognized, and new vessels were outfitted with many of the same instruments. And, for the first time, refrigeration was installed on North Carolina trawlers, particularly those specifically designed for ocean fishing.

Improvements in the net-hoisting gear and the nets themselves were adopted, as well. While many of the trawlers continued to be equipped with power winches and rope towlines, the new, larger vessels were being equipped with steel cable and drum hoists. The cables operated from the main engine. And try nets became more common, particularly on the central coast. In 1948, the "New Jersey" or balloon net was introduced in the Southport area. Because of its efficiency, this net technology rapidly replaced the old flat nets.

Chapter Four

The larger boats, with their diesel engines and advanced technology, had an immediate, significant impact on the shrimping industry on the southern part of the coast. For the first time, they were operating far from home and staying at sea for more than a day. On November 28, 1945, the *State Port Pilot* reported that, while the number of trawlers working in the area numbered only sixty-five, "the production has already broken that of any remembered previous season, even during the years when there were around 200 boats." Nearly three million pounds were landed in the Southport area alone in 1945.

Based on personal interviews, and a search of newspaper articles and archival materials, Bill Still became convinced that many fishermen believed productive shrimping grounds existed offshore, like those found off of the Louisiana coast as a result of government research. They began to discuss the possibility of a similar project in North Carolina. Mr. W.B. Keziah, a well-known supporter of the fishing industry in Brunswick County, wrote to the director of Conservation and Development in April 1945. He urged that government-sponsored exploration be undertaken, the absence of which, he feared, would mean that North Carolina fishermen would move their large boats to known offshore fishing grounds, as in Louisiana. The North Carolina General Assembly created a five-person Shrimp Commission and appropriated $50,000 to finance shrimp-industry research, which included exploration for new harvest areas. The survey, completed in 1945, uncovered no new important offshore shrimping grounds. A good account of this exercise was reported in the Carteret County *News Times* on October 11, 1955.

Although no new shrimping grounds were discovered, the industry continued to spread geographically. According to our interviews with fishery managers and fishermen, the first of the two principal harvest areas were the coastal waters of the mouth of the Cape Fear River, with Southport as the base of operations. At this time, Garland Varnam opened the first shrimp handling and dealership in Southport—Garland's Seafood. It still operates today, with his son, Nick, running the operation.

The fishing grounds extended about ten miles to the west from Cape Fear Point. Some scattered shrimping occurred southward to Little River Inlet, at the South Carolina border. The second principal harvest area (and their northern limit) consisted of the inside waters near the mouths of the Neuse and Newport Rivers, in Core and Pamlico Sounds, and in the nearshore waters a short distance outside of Beaufort Inlet on the central coast.

While the distance of travel and variety of fishing afforded by the larger vessels created more economic opportunities, there were new sets of problems, as well. Interstate rivalries emerged as the large-boat operators sought shrimp and finfish in waters far from their home ports. In some cases, disputes were of such magnitude that, in 1951, the federal judicial system had to get involved. South Carolina had imposed a special tax on non-resident fishermen, but the U.S. federal court ruled against it. During the same period, in which the North Carolina fishermen had reacted angrily to such a tax, they insisted that their own Board of Conservation and Development impose restrictions on out-of-state shrimpers. In its stead, the board passed a resolution extending reciprocal privileges to fishermen from states that permitted fair access to North Carolinians in their waters (Biennial Report of the Department, 1954, Raleigh, North Carolina). And, if there is one thing that I have seen remain constant in my nearly thirty years of resource management research, it is the ever-present attempt of fishermen in *every* state to limit the participation of those from other states in their home territories, while objecting to any effort to prevent themselves from fishing anywhere they wish.

THE POST-WAR BOOM IN PAMLICO SOUND The most important expansion of the shrimping industry in North Carolina waters in the post-war years was the opening up of the Pamlico Sound region in 1951. In 1934, the Board of Conservation and Development had severely curtailed trawling in the sound in order to protect small finfish. The regulations were eased later, and in 1937 some limited shrimping was permitted. "Inside" (or estuarine) shrimping became popular during the war because of the limitations that had been imposed on ocean fishing. Pamlico Sound became increasing attractive to shrimpers because of its convenience and location, its abundance of marine resources, and the safety and comfort of inside fishing. A resource manager we interviewed in the eighties estimated that between 1941 and '49, at least half of the shrimp offloaded at the docks came from the sound and its tributaries.

During those same years, shrimping became, for the first time, the most important fishery in Hyde, Pamlico, and Beaufort Counties, all inland. In September 1948, the Raleigh *News and Observer* wrote about fishing in the sound and declared that the "biggest fishing fleet that ever assembled ... is busy dragging the bottom to cash in on the shrimp bonanza, and it's getting hard to buy a trout or a mackerel or a bluefish in the fish houses." The paper noted that many of the five hundred boats trawling there ordinarily would be fishing for finfish. "But, at 32 cents a pound, the price the ... boat operators pay the fishermen, the lure of the shrimp is hard to resist." Some of the vessels had come from Florida, and a boom-type of mentality had developed in the sleepy coastal littorals on the shores of the sound: "Engelhard, for instance, is taking in about $75,000 a day for a town more used to $750." (See, also, the Lake Landing History, in *Hyde County History*, published there in 1976. As a special note, the town of Engelhard was hit especially hard by Hurricane Isabel in 2003, and its future as a fishing village is in doubt.) And, as I have noted, in 1951, shrimping received a further boost when night harvesting was allowed in Pamlico Sound for the first time.

The rapid acceleration of the industry in North Carolina was accompanied by a new round of state legislative initiatives to regulate it. Prior to 1946, some rules had been put into place by the Board of Conservation, which had the responsibility to manage the state's commercial fisheries, and rarely were they challenged. The shrimping boom changed this. From 1946 on, the state attempted to impose more regulations, which generated considerable controversy. Confrontations began to develop among all the stakeholders—shrimpers, finfishermen, resource managers, and scientists.

Controversies among just the fishermen gained a new dimension in 1951 when the Board agreed to allow night shrimping. The permitted period extended from June 1 to July 15. This decision was based on a recommendation from the Institute of Fisheries Research in Morehead City, which had discovered that spotted (pink) shrimp moved only at night in the sounds. After the policy was put into effect, production in the Pamlico Sound area rose over 200% in the first season alone, according to state statistics.

But opposition to night shrimping died slowly. Many finfishermen continued to oppose it for years to come, blaming the decision for the subsequent declines in various species of fish in certain areas. Nevertheless, the expansion of the industry during this period resulted in a significant increase in the number of vessels engaged in shrimping. In 1946, for example, 289 boats were registered for shrimping; in 1947 that number increased to nearly 500, according to state registration data.

INITIAL ATTEMPTS AT ORGANIZED POLITICAL ACTION The post-war boom in the industry did not bring about consistent prosperity for the fishermen. The prices they received from the fish houses continued to fluctuate, depending upon supply and demand. Fishermen accused the larger dealers of paying too little for the product. Dealers countered that their expenses, particularly labor, were higher. That had been affected by the clamping down on the use of child labor after the war by the U.S. Department of Labor.

Dissatisfaction with the prices fishermen were receiving at the docks led to a strike in the Broad Creek and Swansboro area (west of Morehead City). It also resulted in shrimpers becoming more attracted to organized labor, which was gaining momentum nationwide. In 1947 nearly a hundred fishermen residing in Harkers Island and Atlantic joined the

International Fishermen and Allied Workers of America union, an affiliate of the Congress of Industrial Organizations (CIO). The *State Port Pilot* reported that nearly 75% of the owners, captains, and crewmembers in their area had also joined the union. Newspaper reports all along the coast reported that union organizers from the United Marine Division of the International Longshoreman's Association had attempted to persuade shrimpers in Southport and Pamlico Sound to join their organizations. In 1950 dealers had urged that a fishermen's association be formed, which would represent both the fishermen and dealers in order to stabilize prices and serve as a forum for legislative action. The North Carolina Fisheries Association was finally organized in November of 1951.

THE BASIS OF STATEWIDE MANAGEMENT Industry trends that began after the war continued throughout the fifties and created a number of unresolved issues. Shrimp production, as indicated previously, reached a record of over 14 million pounds in 1953, and by 1955, more than two thousand North Carolinians were employed in the industry in various capacities. Some twelve hundred vessels of all sizes and types were in use, and their number was increasing rapidly. By the end of the century, however, the total number of North Carolina vessels in the industry had dropped to 933. About forty of these were small, inland-based boats of twenty feet or less. About seven hundred were trawlers used exclusively for "inside" shrimping, which left only 219 engaged in ocean harvesting. Over half of the ocean-going vessels were harbored in the southern coastal region, with about a fifth residing in Carteret County (NCDMF, March 1999). The economic implications of this trend are explored in Chapters Five, Six, and Eight.

An important indicator of the real and anticipated trends in the industry in the middle of the twentieth century was the development of the Morehead City Shipbuilding Corporation. By 1956, it had become the largest fishing-vessel shipyard in the southeast. It mass-produced "Hatteras Trawlers," ranging from fifty to seventy-three feet in length. In

1957 the shipyard was launching six or more boats per week. At its peak, the yard employed 250 workers, but in 1959, the company went bankrupt. The collapse occurred primarily because of a credit plan that allowed fishermen to purchase trawlers for a very nominal down payment and then pay off the balance over time. Unfortunately, a series of poor harvest seasons resulted in numerous payment defaults. The literature on this corporation is extensive, especially regarding its aggressive advertising campaigns. The *Southern Fisherman* and other trade magazines carried numerous articles and announcements about the yard and the trawlers. Newspapers in the state also gave much publicity to the venture (Carteret County *News Times* and *The State*, 1959–60). At one point, the shipyard was the largest firm of any type in the county.

Rapid, uncontrolled expansion of the fishery overwhelmed the management capacity of the local authorities and the fishermen and dealers themselves. The increasing number of fishermen competing for a limited and unpredictable resource led to the escalation to controversy and conflict. Legislative reaction reflected a concern over the threat to all marine resources. Statutes were introduced to eliminate the prevailing chaos and restore order. The manner in which this occurred is the subject of the next chapter.

Management: From Local Customs to Statewide Regulations

My colleague Raoul Anderson wrote that there probably isn't such a thing as an unregulated fishery (1979). Prior to the twentieth century most fishing activity was controlled by custom and traditions, either for conservation purposes, the control of wealth, or both. This has changed worldwide, however. Disputes over resource allocation that cannot be decided locally, along with declining resources, have stimulated more and more formal regulations by state, regional, national, and international bodies. In 1982, my colleague, Mike Orbach and I used the term "upward aggregation of power" to describe the ramping-up process by which local control gives way to large and distant management regimes. More often than not, the consequences of such a change are far more extensive than those who initiated the process bargained for. Issues are defined differently than originally presented, often requiring control over more local, as well as extra-local activities, than those involved had thought would be the case. In some cases, things spin out of control, with lots and lots of unintended consequences.

With regard to North Carolina's development of a management framework for the shrimp fishery, particularly in the crucible of the tumultuous 1960s, the dominant issue became the growing economic and political importance of fishing versus the use of the coastal areas in which fishing occurs for other purposes, such as shipping, retirement living, and recreation (Maiolo and Tschetter, 1982). By the 1980s, the state had made the transition from custom to law in the regulation of most of its fisheries. As the value of shrimp became increasingly clear, this general trend was reflected in carefully designed efforts to regulate its harvest and sale. Control was no longer left up to the fishermen and dealers. For example, when channel-netting began to interfere with the efforts to trawl in the same waters, state law was developed in a way to preempt local customs, providing the trawlers respected the space of the channel-netters. The issue could not be settled among the fishermen alone.

The development of the state's shrimp fishery from the turbulent sixties to the present is best understood by examining the comprehensive management framework that began to take a definite shape just as President John F. Kennedy took office at the national level and Terry Sanford became governor of North Carolina.

MANAGING COMPETITION AND CONFLICT[1] Initial efforts to regulate the fishing industry in North Carolina extend back to the eighteenth century. From 1784 to the present, the General Assembly has played a role through the process of creating *general* statutes. In the latter part of the nineteenth century, the state created the first agency to manage fisheries, but it was not until 1915 that a state agency, the Fisheries Commission, began to develop policies specific to shrimping. Since then, the state has attempted to improve shrimp harvests, to increase economic returns, and to handle the increasing number of disputes in the fishery. But, until the 1960s, these policies had been generally guided by public hearings, petitions, local interest, and political pressure. Beginning in the sixties, policies and regulations began to reflect the incorporation of scientific research into the management of the fishery. In October 1963, Governor Terry Sanford announced:

> The basic function of the Division of Commercial Fisheries of the Department of Conservation and Development has been the enforcement of laws relating to the commercial fishing industry....

Chapter Five

Although law enforcement is, and will continue to be, an important activity of the Division of Commercial Fisheries, it must be supplemented by an expanded program of research .... (*Biennial Report of the North Carolina Dept. of Conservation and Development, 1964*).

This basic change in philosophy would affect future management decisions about the fishery. The change, to a great extent, was due to a bitter struggle between commercial and sportfishermen. Recreational fishing had been expanding rapidly since the end of the war. It had been estimated that, by 1959, there were more than 328,000 saltwater anglers in the state alone. State Senators S.B. Frink and Cicero Yow introduced a bill that would repeal the Bureau of Conservation and Development policies prohibiting commercial activity on weekends, which, in turn, would permit *recreational* shrimping, that is, the use of commercial gear by recreational fishermen. The bill aroused the anger of commercial fishing interests on the board but was adopted in May. Although the bill involved only three counties and concerned only weekend shrimping (along with recreational summer oystering), it was symptomatic of the tensions between commercial and sportfishing interests in the state and elsewhere.

More recreational pressure emerged in October 1960 when, at the insistence of sportfishing interests, the Conservation and Development board agreed to a hearing on a proposal to forbid trawlers from operating within the three-mile territorial limit from Cape Hatteras in the north to Cape Fear in the south and in some parts of Pamlico Sound. Sportfishermen had complained for a number of years that trawlers operating close to the beaches and piers were depleting the supply of fish (Raleigh *News and Observer*, October 18, 1960). A special committee was created to consider the problem and recommend action to the state board. That committee decided against the proposed trawling restrictions for two reasons. First, activity would be limited in too much territory. And second, which was a breakthrough reason, there was not enough scientific information on which to base an informed decision (Raleigh *News and Observer*, January 10 and April 8, 1961).

Management: From Local Customs to Statewide Regulations

In June 1961, a unique bill was cobbled together and introduced into the General Assembly. It included the regulation of shrimping on a statewide basis. In effect the bill would address the nagging problem of permitting *local* interests to regulate local fisheries by *local* ordinances. The bill had the support of Governor Sanford who, in August 1961, wrote, "Local laws make exceptions for local interests. Commercial fishing is a statewide industry and should be protected and developed on a statewide basis. Local legislation can dilute and threaten the effectiveness [of statewide programs]" (Governor Sanford papers, North Carolina State Archives, Box 36). The governor had asked for and received considerable support from commercial fishermen for this change, but sportfishing interests were less than enthusiastic (Raleigh *News and Observer*, June 8, 1961).

The increasing tension between the two fishing groups prompted Governor Sanford to consider reorganizing the Bureau of Conservation and Development's Division of Commercial Fisheries. One option was to completely separate the two, but the commercial fishermen were divided on the matter. Wanting some form of change, the governor created a study commission. By 1964, the commission made its restructuring recommendation, after first criticizing legislation that had been passed to appease interest groups, political appointments to key resource-management positions, and confusion over policy interpretation. The "inefficiency" that was produced could be overcome by first abolishing all local statutes regulating commercial fishing.

The North Carolina Fisheries Association (NCFA), principally representing the large fishing and seafood-processing interests, wrote to Governor Moore opposing the commission's restructuring recommendation of the study commission. They did inform the commission, however, that they did not object to having sportfishermen on the division's advisory board (Raleigh *News and Observer*, May 12, 1965).

In June the General Assembly passed a new fisheries act. Effective January 1, 1966, the Division of Commercial Fisheries would become the

Division of Commercial and Sport Fisheries and would have an Advisory Board. The legislature also accepted the commission's recommendation to disallow local statutes. Fishing laws were to be rewritten with the primary objective of applying policies on a statewide basis. For example, a statute permitting night shrimping would have to apply to the entire state, not just to selected counties or fishing areas. Additionally, the new legislation recognized the need for continuing scientific research in fisheries management.

The evolution of the state's fisheries-management agencies did not end with the 1966 legislation. More lessons were learned, and administrative and political exigencies created a momentum for further refinement. In 1971 the General Assembly completely reorganized state government. The Department of Natural and Economic Resources was created with the mandate of managing the state's fisheries. In July of that year, the Commercial and Sports Fisheries Advisory Board was transferred to this agency. And later, in 1975, under Governor Holshouser and the Executive Organization Act, the North Carolina Division of Marine Fisheries, which still exists today, was established. This accomplished what sportsfishermen had been urging for years: elimination of the distinction between "sports" and "commercial" in the Division's title.

Between 1963, when Governor Sanford stressed that fisheries management had to develop a new philosophy based on research, and the enactment of the reorganization legislation, resource management underwent a significant theoretical change. A report on the history of the Division of Commercial and Sports Fisheries noted that no longer would "regulation and management of coastal fisheries ... [be] ... accomplished through political action, based on opinion and emotion, but on fact, and facts must be obtained scientifically" (Governor Moore papers, Box 17, North Carolina State Archives). With nearly thirty years experience with the state's fishery-management system, I can say that there is a great deal of truth to this statement. However, it would be unrealistic to imagine that politics did not enter into the decision. They did, and still do. And, in a democratic system of government, why should we expect anything less? In fact, one could make the argument that decisions based exclusively on scientific evidence, and to the exclusion of politics, are as dangerous as those based exclusively on political considerations. What has impressed me about resource-management decisions generally, and specifically those in North Carolina, is that they eventually tend to present a pretty good blend of the scientific and political.

TRASH FISH AND TRAWLING REGULATIONS The changes made by the state in fisheries management in the mid-sixties did not end the controversies and confrontations between sport and commercial fishing interests. Two issues that were particularly sensitive in the years following were trash fishing and trawling. Both issues had long histories and, in fact, were factors in the reorganization of the state's fishery agency.

North Carolina has the largest estuarine system of any state on the Atlantic coast, with over two million acres of open water, and nearly two hundred thousand acres of wetlands. The system consists of shallow water, with extensive sea grass beds that serve as nurseries for 90% of the state's commercially important marine species. By the turn of the century (2000), about 46% of the estuarine waters had been closed to trawling in order to protect both the habitat and the juvenile animals that dwell there. Some argue this is not enough, while others plead the case for fewer restrictions. The debate does not come to a halt at the inlets, but continues on out into the near-shore ocean, also a habitat for young fish. The controversy has left an interesting mosaic of clashes among various interest groups, management systems, and the managers themselves. Let's look at the picture more closely.

Trash Fishing "Trash fish" are so called because they are too small to market (small trout) or not marketable for human consumption (menhaden). For years commercial fishermen had captured these fish in the bays and the near-shore ocean and then sold them for meal, oil, fertilizer, and crab bait. The return on the sale of these products was rarely sufficient to support fishing operations for these products alone, but it did help offset costs for harvesting targeted animals. Also, permission to allow possession of such fish prevented the dumping of large quantities of small dead and dying fish inshore and inside the state's inlets.

In the 1950s, the Conservation and Development Board attempted to address the problem by putting limits on the amount of total catch that could be sold to reduction plants (for the production of oil, fertilizer, etc.). Sportfishing interests would have none of that. The 1963 Commercial Studies Commission recommended the complete elimination of trash fishing but did not suggest a method to achieve that goal (Raleigh *News and Observer*, March 31, 1965). Research conducted by the University of North Carolina Institute of Fisheries Research provided a rationale to put the 25% limit into effect, along with targeting juveniles (Raleigh *News and Observer*, January 21, 1966).

Eventually, the limits were expressed in terms of number of boxes of trash fish per total catch. With regard to the capture of trash fish outside of the inlets, not much has changed since that period. Mesh sizes and box limits prevail to this day.

Trawling Trawling presents a different set of problems, however. In addition to habitat destruction in the estuaries and the capture of juveniles in the near-shore waters, trawls operating along the beach interfere with sportfishing on the piers. Since the beginning of World War II there had been considerable opposition to the use of trawls in inshore, as well as inland waters. Opponents have wanted to restrict trawling within one mile of the state's beaches.

Many rules have been put into place regarding trawling in these internal waters. There are now limitations on incidental catches of finfish, and these limitations change as the shrimping season comes and goes. The director of the DMF has considerable latitude in closing or opening specific areas for trawling in order to protect juvenile shrimp and "undersized" juveniles of other species (NCDMF, March 1999: 40–41). But, in regard to the shrimping industry specifically, the incidental or bycatch of species other than shrimp, when shrimp are the target, has received a great deal of attention in the state during the past few years. Further, "while ... bycatch ... has been a concern ... since the 1950s only recently has the affect of bycatch mortality been examined, and only for two species, weakfish (trout) and Atlantic croaker .... The bycatch of weakfish in the shrimp trawl fishery has been the major source of mortality for this species" (DMF, March 1999:20). It is also the case that two other species, namely, spot and southern flounder have been impacted, although the latter has been more affected by crab than shrimp trawls (DMF, March 1999: iii).

In 1992, the DMF, based on its own in-house research and fishery management efforts at the regional level, put into place a requirement mandating that shrimp trawlers working in state waters equip their nets with functional fish excluders or Bycatch Reduction Devices (BRDs). These allow small fish to escape from the net before reaching the tail bag. North Carolina was the first to require such devices (DMF, March 1999:21). Since that time, the Division staff has worked with commercial fishermen to find the technology that would minimize the loss of shrimp while also minimizing finfish bycatch. By September 1997, through Proclamation SH-9-97, shrimp trawlers *had to be equipped* with "one of four approved BRDs" (DMF, March 1999:21.

LICENSING AND THE MIRACLE OF 1997 The use of commercial gear and the sale of seafood have been the subjects of debate for decades. Recreational fishermen claim that unlimited trawling, with non-discriminating nets, clam and scallop dredging—virtually all of it—take too much of the resource and damage the environment. Yet, historically, thousands of recreational fishermen, as many as 14,600 in 1982, had licenses to use commercial gear during the 1980s. Many recreational fishermen, who pushed hard and successfully for regulations that limited commercial catches of king mackerel during the eighties, would sell their own catches to offset their sportfishing activities. This was also the case for vacationers at the coast, who would purchase a shellfish license to rake for clams and then sell a portion of their harvest.

Commercial fishermen feel that catching fish for sale is their exclusive right. Further, there is a great deal of support for regulations *for the other guy*, who is not always just the recreational fisherman. The management literature is full of references to efforts to regulate commercial activity by *other commercial fishermen*. In some cases this grows out of competition for space or between commercial fishermen using different types of gear. During the 1980s, New England scallopers, who used dredges, and North Carolina shrimpers, who fished with trawl nets during certain periods of the year, wound up in a battle over the resource. I was involved in the research on the impacts of alternative regulations and watched as the New England Fishery Management Council essentially regulated the North Carolinians out of the fishery by putting size limits on the catches in a way that gave the New England fishermen a competitive advantage. The reason—the North Carolinians out-fished their New England counterparts. In fairness, the North Carolinians got a little greedy and were fishing on grounds that were producing too many undersized scallops.

Within North Carolina itself, as we have seen in this and other chapters, different user groups, as they are called in fisheries parlance, are always trying to get the advantage—which, I suppose, is no different from what often happens in any other work setting. Another of the ways in which the groups are divided, in addition to those I have already discussed, is the way *commercial* fishermen are divided up between *full-* and *part-*timers.

Full-timers are what you pretty much imagine them to be. Most, if not all, of their income comes from fishing. I include the qualifier "most"

Management: From Local Customs to Statewide Regulations

because, historically, many fishermen have engaged in other work activities. Things are changing for some, and I will have more to say about that in the next chapter. Also, with few exceptions, full-timers are "homegrown" on the coast.

The part-timers, however, are not a uniform group. In fact, they are quite diverse. Some are retired "regular" commercial fishermen. Others are what my colleague, Jeff Johnson, labels as those who contribute to the *gentrification* of commercial fishermen. That is, they retire from other occupations, many from the military, and supplement their incomes through catching and selling fish. Still others really have had land-based (and inland) jobs and have been recreational fishermen who widened their efforts to the point that selling fish has become a serious contributor to their incomes. And I have known men who work for the federal or state government and who grew up in fishing families but chose to get a land-based job. Still, they kept their options open and took vacation time to do some serious commercial fishing to supplement their incomes. One of my very best friends on the coast, Manley Gaskill, did this and unfortunately perished one December evening as he pursued his first love, outside of his family: commercial fishing.

The various groups I have described have been going at each other for decades, as we have seen. Much of the regulation I described earlier has come as a consequence of the battles between and among these groups. Often, the attempt to solve one issue would set off convulsions from one fishery to another, from one user group to another, or both.

In the mid-nineties, the campaign to summon the resources necessary for some kind of a permanent solution was ratcheted up to a level that resulted in new regulations that many, including me, believed could never happen. Under intense pressure, legislators from many quarters and weary fishery managers, as they have done in so many instances, began to look for new and creative ways to address the competition issue head on. It appears as if their work during this period will have the same level of long-lasting and significant impact as the changes made during the sixties. This time, the effort mainly, but not entirely, took aim at licensing regulations through the proposed adoption of the dreaded "limited-entry" policy. The concept had been used in a number of fisheries (e.g., the salmon fishery in Alaska) but has been controversial from its inception.

First, we need to cover some background. During the 1980s, the number of commercial fishermen increased from about four to five thousand. The number of part-timers actually decreased from 7,700 to 5,700. After hitting the record number of more than 14,000 in 1982, the number of recreational fishermen with licenses permitting the use of commercial gear declined steadily to under ten thousand in 1990 and dropped under 5,000 in 1992. By 1993, the numbers changed again: the numbers of full-time licenses issued increased to nearly 6,300, part-time commercial licenses to over 6,000, and recreational licenses to more than 9,300, which I will explain below. Competition intensified at the same time fishery managers were trying to come to grips with declining fish stocks. And it was becoming apparent that a number of East Coast fish stocks were in trouble, from New England to Florida. There was increasing fear that, as other states began limiting entry into and effort in a number of fisheries, there would an influx of out-of-state commercial fishermen into North Carolina. This, it was believed, would exacerbate an already distressed situation. During the same period, beginning in the eighties, dramatic increases occurred in the number of entries into the crab fishery. Commercial fishermen sought a freeze (i.e., a moratorium) on new entries, the first step toward a limited-entry policy. The intersection of these events led to major changes in the management of the state's living marine resources.

Legislative leaders became supportive of the idea of a moratorium that grew out of the crisis in the blue-crab fishery and expanded the initiative to all commercial fishing. During the 1993/94 legislative session, the North Carolina General Assembly—on the recommendation of the DMF chief, Dr. Bill Hogarth, and the Marine Fisheries Commission—

established a Moratorium Steering Committee (MSC) to oversee a study of the entire management system and to recommend needed changes. Mr. Bob Lucas—an attorney from Selma, North Carolina, and chair of the Marine Fisheries Commission—was appointed by Governor Hunt to Chair the MSC, as well. Representatives of the various sportfishing and commercial interests, along with scientists, dealer/processors, and environmentalists were appointed to the nineteen-member committee. Five subcommittees were named: Gear, Habitat, Law Enforcement, Marine Fisheries Commission and DMF Organization, and, the subject of our discussion here, Licensing.

The legislature did, indeed, enact a two-year moratorium, later extended to three years, which became effective on July 1, 1994, and extended to June 30, 1997. In early 1997, when it became apparent that some form of limitations on licenses was going to be made permanent (i.e., a limited-entry policy), another extension was passed by the legislature in order to buy even more time.

What happened during this period can be difficult to follow, and I am grateful to Dee Lupton, Janice Fulcher, and Nancy Fish, all of the North Carolina DMF, for helping me get it straight. The first change from the traditional licensing system came in the early nineties with the creation of a vessel decal. This meant that any vessel using commercial gear needed to display the decal, the cost of which was indexed to the length of the vessel. Under this regulation, boats with the decal could use any gear types except those used for certain species of shellfish (i.e., clams, oysters, crabs, and scallops).

An Endorsement to Sell (ETS) provision was added to the license requirements beginning January 1, 1994. This meant that, from that point forward, seafood caught with commercial gear could only be sold by those who had an ETS. This regulation grew mainly out of a growing concern for the need to prevent recreational fishermen from increasing their efforts and selling their catches. This was quite a change because, as I noted, before this time a fisherman only needed the vessel decal in order to use commercial gear and sell the resulting catch.[2]

By July 1994, the moratorium was up and running. A month earlier, the news had spread like wildfire that the moratorium on commercial fishing licenses was about to be put in place. Dr. Hogarth had announced this to the DMF staff, giving the justification that some restrictions were necessary to give his managers time to get a grip on managing some of the species that were being stressed (e.g., flounder and trout). Once this rumor got traction, according to the DMF license people, there was a huge run on boat decals by part-timers and recreational fishermen, which created a spike in license sales.

According to the Joint Commission, "The license moratorium was expressly created in light of ... the duty to preserve and protect its ... resources" and, among other things, "economic turmoil in the commercial fishing industry ...." (Fisheries Moratorium Steering Committee, 1996:viii). It also took into account the historical importance of commercial and recreational fishing, and the possibility of a federal takeover if the state did not get its own house in order. The report referred to "substantial shortcomings in the State's traditional, regulatory fisheries management system" (Fisheries Moratorium Steering Committee, 1996:viii).

The Committee met regularly from November 1994 to August 1996. In July 1995, 225 thousand dollars were allocated to conduct a series of studies on a range of topics. Regular monthly meetings were held for the entire committee, in addition to dozens of sessions held by the various subcommittees. And between August and September 1996, nineteen public hearings were held in locations throughout the state, not just on the coast, to solicit public input on the draft recommendations that had been developed.

Rumors, news articles, telephone calls, and letters to the committee were plentiful, as the various fishing groups and their representatives tried to come to grips with what was happening. As you can imagine, some of the exchanges were heated. Testimony from part-time commercial fishermen complained that limits on gear that would come from one type of proposed license to which they might be subjected would seriously impact their annual incomes. Small-scale operators complained that the proposed rules would favor large operators and dealers. Interestingly, many recreational fishermen spoke out in favor of the moratorium and impending license restrictions (e.g., *Cypress Group News*, February 13, 1996). In July, a memo from the North Carolina Fisheries Association indicated "steadfast" support for the moratorium and many of the developing recommendations of the committee, but it came out against the limited entry licensing proposals under review. The president of the organization, however, later indicated what seemed to be an important endorsement for regulations preventing recreational fishermen from having a commercial license (Raleigh *News and Observer*, August 28, 1997).

In preparation for the public hearings, Chairman Lucas mailed out a letter to those he considered to be interested parties throughout the state, first urging them to attend, and then to help distribute the news about the hearings. Attached was a brochure entitled "It's Up to You," indicating when and where the meetings would be held, the background of the committee's formation, and a summary of recommendations emanating from the subcommittees' work. Mr. Lucas emphasized the "public role" in the formation of the work and the need to stay involved as the process came to a conclusion. Reports from the hearings indicated that, with some exceptions, a consensus was developing among the various parties interested in the proposals the MSC was examining.

A "Final Report" was adopted in October 1996 and presented to the Joint Legislative Commission on the twelfth of that month. The thrust of the report was that "piecemeal fisheries regulation does not adequately conserve, protect or allocate North Carolina marine and estuarine resources." In order to "fix" the problem, the committee proposed a three-tiered licensing system (Fisheries Moratorium Steering Committee,

1996:ii). On August 14, 1997, Governor James Hunt signed into law the Fisheries Reform Act, the centerpiece of which was the licensing system recommended by the commission, albeit with some changes. To become effective on July 1, 1999, the Act first continued the existing moratorium. And under the new system, only fishermen holding the Endorsement to Sell license would be eligible to purchase a new commercial license. The change that became effective was the placement of the Vessel Endorsement to Sell under the moratorium. All existing licenses could be renewed until the new system went into effect in 1999.

The first of a new, multi-tiered system was the Standard Commercial Fishing License (SCFL, pronounced "scuffle"). This license is for commercial fishermen harvesting shrimp, finfish, crabs, and other shellfish, and costs $200 annually for residents, and $800 annually for non-residents. The licenses are tied to an individual, like a driver's license, not to a business. A free Shellfish Endorsement accompanies the SCFL, if it is requested.

Only fishermen who had held a valid ETS on July 1, 1999, were eligible to purchase the new license, thereby "grandfathering" their privilege. Vessel-endorsement fees, part of the preferred Commercial Fishing Vessel Registrations, are required for the use of boats to harvest, and these fees are indexed to vessel length. The ETS requirement disqualified a number of part-time fishermen who had failed to spend the money to buy one earlier. And another interesting twist surfaced, as well. While no new, additional entrants were allowed into the commercial fishing industry, a fisherman was allowed to transfer and even sell his or her license under a specified set of rules. And, while a cap was placed on the number of licenses on July 1 (8,400), provision was made for an additional five hundred SCFLs to be allocated according to pre-established criteria, the main ones being past involvement in commercial fishing and dependence on fishing for a living. This brought the allocation to 8,896. The number of SCFLs actually issued (including R-SCFLs, discussed below) dropped by about 2% over the first three years of the new program.

Management: From Local Customs to Statewide Regulations

Apparently, a number of part-timers and even recreational fishermen who had been supplementing their incomes with impressive amounts of fish sales had not been filing such information in a way that would have substantiated their dependence on fishing. And since they had not spent the money for an ETS before the moratorium was put in place, they would have to petition a three-member board successfully in order to obtain an SCFL under the new system. The criteria require either proof of a history of commercial fishing in their families (for young entrants) and/or proof of economic dependence on fishing in three of the last five years or for three years prior to the past five. Since many of them could not qualify, they were left to either purchase an SCFL from someone else or use a Recreational Commercial Fishing Gear License (see below) and continue what they had been doing previously. The problem with this was that it introduced even more restrictive gear rules, along with the danger of being caught selling fish illegally. We will return to this problem presently.

Special provisions were made for retired commercial fishermen who still have some interest in and/or economic dependence on fishing. The special license for such individuals is called a Retired Standard Commercial Fishing License (R-SCFL—pronounced "are-scuffle"). To qualify, a fisherman has to be sixty-five or older, already hold an ETS, and pay a modest $100 annual fee ($800 for non-residents). It is not assignable and requires a Vessel Endorsement decal (CFVR) for any boat using commercial gear. The number of these licenses has increased between its inception (about 500) and 2002 (676). And indications are that the number will increase again in the coming years.

A Recreational Commercial Fishing Gear License, the (RCGL—pronounced "ruckle"), allows recreational fishermen to use a limited range of commercial gear to catch fish for personal consumption only and not for sale. As in the past, one of the most open secrets in the seafood industry is the violation of the not-for-sale provisions in the licensing of recreational fishermen to use commercial gear. The cost of the RCGL is $35 for residents and $250 for non-residents. Like the R-SCFLs, these are not assignable, and those who use them are subject to certain size and bag limits. During the first three years of the new program, the number of active licenses issued in this category seems to have dropped from just under 6,800 to 5,400 in 2002. But it is hard to tell exactly what this means because a person can replace a lost permit and it will show up as an addition instead of a replacement. According to Dee Lupton of the NCDMF, 5,400 more accurately reflects the number of individual license holders.

Regulations that had already been adopted in 1994 were included in the Act to limit the number of licensed dealers. Under the Act, dealers continue to be required to buy seafood only from licensed commercial fishermen. Properly licensed recreational fishermen can catch seafood with commercial gear, but not sell it. Only residents can buy these licenses, and there are separate fees for different species. There is, however, a "consolidated," all-species license that costs $300. The number of licenses issued in this category has remained constant at about 830.

A flurry of activity, in the best sense of democratic involvement, occurred between October 1996 and the adoption of the Act in August 1997. But Marine Advisory Board meetings were filled with a great deal of animosity. There seemed to be a collective sense that time-honored traditions governing the use of commercial gear and the ability to sell harvests to supplement part-timers' incomes were begin trampled on. Phil Walker, president of the North Carolina Fishing Freedom Fighters complained, "This all started because some crabbers said there were too many pots and not enough crabs.... But we had a bumper year for crabs. But they continue to tell us the sky is falling." He then went on to accuse the policymakers of developing new regulations to justify their jobs and pander to special interests.

The Joint Legislative Committee's scheduled votes on the proposals were delayed in early 1997 for several months because of the complexity of the reforms and the concerns still being expressed by various interest groups. A key player, State Representative Redwine, who would eventually co-sponsor the Reform Act, argued that more time was needed to muster public support for the reforms. Mr. Lucas countered that the existing proposals already included public input and exhibited a sufficient public consensus (Carteret County *News Times*, December 11, 1996).

By July 1997, many of the remaining issues had either become better understood or been resolved. The Senate Committee on Agriculture, Environment and Natural resources passed the bill after eliminating a proposal for a recreational saltwater license (Raleigh *News and Observer*, July 23, 1997). The next stop was the Finance Committee and then on to the full House and Senate chambers. On August 13, all indications were the Act would sail through both chambers. On August 14, the bill was passed unanimously in the Senate, but barely in the House. A last-ditch attempt by three representatives from Brunswick County to stall acceptance of the Act failed, but their efforts changed an expected 98–7 vote to 56–49. The entire Onslow/Carteret delegation except State Representative Jean Preston voted against the Act. Their concern was that more than 1,800 fishermen would lose their commercial fishing licenses because they did not hold an Endorsement to Sell. They perceived this as unfair and called for a grace period to allow for a review and resolution of the problem(*Jacksonville Daily News*, August 15, 1997).

Governor Hunt, who had endorsed the reforms from the beginning, immediately signed the bill into law, scheduling the reforms to go into effect on July 1, 1999. It really was a remarkable achievement, especially since it had not been given much of a chance for success by many that had seen more than a few attempts at extensive reforms, especially of licensing policy, fail. Before I go farther with this, permit me a moment to give my personal kudos to those involved in the passage of the Act. Those who have not been involved in fisheries management may not appreciate the significance of this accomplishment. In virtually all of my past experience with the adoption of new regulations, management agencies would usually open the bidding by declaring, "You can't fish anymore" (I am, of course, being a bit facetious). To which the fishermen would respond, "No deal; we fish all we want, but our competitors cannot fish at all!" And then the negotiations would begin. While some good and useful legislation might result, and it sometimes did, too often the result has been some sort of a compromise that made little sense to anyone—the proverbial "camel," if you will. The 1997 Act distinguished itself by first declaring a moratorium, establishing the criteria for using commercial gear and selling seafood, and then, even though some minor compromises were made, *sticking to the basic mission*. The State of North Carolina is indebted to the courageous men and women who made this happen.

As I indicated earlier, the scope of the Act is seen in its incorporation of other provisions that appreciably altered many of the existing management practices, including: 1) reduction of the Marine Fisheries Commission from seventeen to nine members, with very specific provisions on representation; 2) mandating development of Fishery Management Plans (FMPs) for all of the state's commercially and recreationally significant fisheries (including one for shrimp, currently in development); 3) establishment of a joint effort with other state agencies to develop Coastal Habitat Protection Plans; 4) provisions that provide for stiffer penalties for law violators; and 5) much to the delight of scientists, especially social scientists, a mandate to conduct a series of important studies, including, among others, feasibility studies on additional licensing and the management of shellfish.

With that discussion behind us, we are now up to date (as of the late months of 2003) with the statewide management system. It has been quite a journey since the first fishery regulations were put in place in the late-nineteenth century and the first set of regulations directed toward shrimping. Now let's take a look at this "Commercial Man," as he calls himself. You know whom I mean. He's the one who gets up early in the morning and sometimes finishes late at night so we can enjoy all of those tasty foods with that wonderful omega-three fish oil.

The Commercial Man[1]

Fishermen have to adapt to variations in the availability of their resource targets and the public's demand for them. Both change seasonally and over longer periods of time. *Normal fluctuations* result in accommodations to *cyclical* resource, capital, or market events, and *relatively predictable changes* in activities, which are repeated in successive annual cycles. *Long-term directional changes* occur because of new resource discoveries, stock declines, technological innovations, or changes in the marketplace. A good example of this is the manner in which shrimping became embedded in the larger seafood industry during the 1930s. This chapter examines each of these types changes.

ADAPTATIONS TO GROWTH AND CYCLES Marine-resource cycles are very normal in North Carolina. An examination of historical fluctuations in the shrimp industry illustrates this. Table 6.1 shows the fluctuations in shrimp landings in selected years since 1967 (the base year for the overall Consumer Price Index). The range extends to over six million pounds. Ex-vessel values (the price fishermen receive at the dock when they sell to dealers) have grown to thirteen times their 1967 value! These numbers are not sufficient, however, to measure all trends in the fishery. Changes in resource abundance, which, in turn, affect the landings, do not adequately reflect changes in economic circumstances, especially the influence of inflation. A more comparable means of examining the data is to "deflate" the changing values, i.e., adjust for inflation. "Constant" dollars provide a better measure of what economists refer to as "real" growth or decline. Fishermen's reactions to real growth or decline may explain why adjustments in fishing activities do or do not always occur in the direction that might be expected on the basis of reported statistics.

Table 6.1 illustrates the differences between adjusted and unadjusted variations in shrimp ex-vessel sales (value) during the latter decades of the twentieth century. The examination begins with 1967 for three reasons: First, U.S. government data for U.S. cost-of-living changes use 1967 as their base year. Second, at about that time, shrimp-fishery management efforts in North Carolina were intensified. Third, the period from 1967 to the present shows an impressive record of increases in unadjusted value.

As I said, the U.S. government uses 1967 as its baseline year (100%) for the Consumer Price Index (CPI) for all consumer items at the national level. Each subsequent year's adjusted value for each consumer item is stated in 1967 dollars.[2] Note, in Table 6.1, the incredible increase in the value ex-vessel sales up through 2000, from 1.8 million dollars to nearly 25.405 million. This translates into an increase from 37 cents to $2.91 dollars per pound. But when an inflation adjustment is made by using the CPI for all items, it becomes obvious that the real growth has been substantially less. The adjusted total drops to slightly over five million dollars and the per-pound average plummets to fifty-eight cents.

TABLE 6.1[3] North Carolina Shrimp Landings, Ex-vessel—Total and Per-pound Values. Adjusted per CPI (All Items): 1967–2000

Year	Pounds Landed	Total Value	Value per Pound	CPI (All Items)	Adjusted Total	Adjusted Value per Pound
1967	4,919,000	$1,809,000	$0.37	100.0	$1,809,000	$0.37
1970	5,054,000	$2,493,000	$0.49	116.3	$2,143,594	$0.42
1975	5,164,000	$5,054,000	$0.98	161.2	$3,135,236	$0.61
1980	9,823,000	$17,185,000	$1.75	247.0	$6,957,000	$0.71
1985	11,683,427	$21,130,303	$1.81	318.5	$6,634,317	$0.57
1990	7,839,457	$15,885,027	$2.03	384.4	$4,132,420	$0.41
1995	8,669,100	$20,316,560	$2.34	446.1	$4,554,260	$0.53
2000	10,334,325	$25,404,839	$2.91	499.1	$5,080,968	$0.58

Even more dramatic is the effect when a Producer Price Index (PPI) for Frozen and Packaged Shellfish and Other Seafood is applied and then further refined by adjusting for Unprocessed Shellfish only (also a PPI. Table 6.2). Data for these two indices became available in 1990 (Frozen and Packaged) and 1982 (Unprocessed), respectively. The only starting point I could use for all indices for comparison is 1990. Table 6.2 indicates a real decline in returns to shrimp fishermen, provides an important element in the mosaic of the commercial fisherman's pursuit of his livelihood, and sets the background of the almost vicious competition with others that would threaten that livelihood.

TABLE 6.2 Ex-Vessel Shrimp Landings—Value per Pound, Adjusted by CPI and PPI (prices to producers) Categories: 1990-2000

Year	Unadjusted	Adjusted by CPI (All Items)*	Adjusted by Frozen and Packaged Shellfish and Other Seafood**	Adjusted by Unprocessed Shellfish CPI***
1990	$2.03	$0.53	$1.87	$2.03
1995	$2.34	$0.53	$1.68	$1.93
2000	$2.91	$0.58	$1.73	$1.71
Change 1990-2000	+43%	+9%	-8%	-16%

*Adjusted by 1967 CPI base year. **Adjusted by 1990 PPI base Year. ***Adjusted by 1982 PPI base year

The chief suspect in the real economic decline is the impact of imports, especially from countries that have developed shrimp aquaculture (also called fish farming or fishponds). Data on these imports for 1989 and onward are available from the National Marine Fisheries Service, Fisheries Statistics and Economics Division website (http://st.nmfs.gov/pls/webpls/tradeprodcten-trycum.results). Even before 1989 the data indicate that, following the poor domestic-harvest years in the late seventies and early eighties, imports had increased from modest levels to over 341 million pounds in 1983. Later in the decade, about 500 million pounds were imported in both 1989 and 1990, but the price slipped. For example, the average price per pound in 1989 was $3.40, which slipped to $2.93 in 1990. These are not ex-vessel prices, but wholesale values to distributors in the United States. That means that foreign dealers/processors buying the shrimp in the same fashion as North Carolina dealers are probably paying about one dollar per pound or *less* at the initial point of exchange! The shrimp imported into the United States are headed, some breaded, all peeled, and may have undergone other forms of processing. To be equivalent, all of this processing would at least double or even triple the price, depending on the kind of processing, for North Carolina shrimp values. The impact of the imports, then, can be seen in the tremendous difference in price between domestic and imported shrimp, which holds down the price of the domestic shrimp. Let's take a closer look.

An examination of the entire decade indicates that imports grew to more than 759 million pounds in the year 2000, selling for more than 3.7 billion dollars! The average price per pound for processed imports was $4.87, an unadjusted increase of 43% since 1989, but 66% since 1990. Recall that the unadjusted increase for North Carolina shrimp was 43% and the adjusted change was -8% (frozen and packaged). For the imports that adjusted change also was found to be -8% from 1989, but an increase of 7% over 1990.

By the year 2001, shrimp imports had grown to more than 882 million pounds, but the price dropped to about $4 per pound, a nearly 18% decrease in the *unadjusted* value alone. Shrimp imports accounted for 37% of the value of total edible imports. The range of countries from which the shrimp come is impressive, indeed. Ten nearby Latin American countries, led by Mexico, sent us nearly 147 million pounds; followed by ten countries in South America (the leader was Ecuador) with another 147 million pounds; then contributions by more than two dozen countries in Europe (including the United Kingdom with 291 thousand pounds); and Asia, Oceania, and Africa. Thailand is a huge player in this picture, sending close 300 million pounds, more than one-third of the total (U.S. Dept. of Commerce, September 2002:66).

Competition from imports in the marketplace, then, is worldwide, and has become an important thread in the North Carolina shrimp fisherman's occupational and cultural fabric. Foreign production has continued to increase, and it has appreciably affected the prices of domestic shrimp in general and North Carolina shrimp in particular. If you want to see the effect on prices for yourself, visit one of the discount markets like Sam's Club. Go to the frozen-food section where you will be able to buy two pounds of headed and de-veined small, salad shrimp for $7.88, two pounds of *cooked* 20/24 count cocktail shrimp for $13.88, and larger ones in one-and-a-half-pound bags for $14.98 (from Thailand). Some dealers prefer to buy and distribute imported shrimp because they perceive the quality to be better and more consistent. But this is not the way everyone in the industry looks at the matter, and we will need to revisit the issue again in this book.

Seeking relief from this problem, Florida shrimpers confronted their state legislators. In response, the legislators proposed a job-retraining program for the fishermen, who rejected the proposal out of hand. Instead, they called for the repeal of certain regulations and fuel taxes to reduce their costs. One fisherman blamed part of the problem on the required use of Turtle Excluder Devices in their nets, which he felt drove up the cost of harvesting. The shrimpers also appealed to the lawmakers to put tariffs on imports. The North Carolina Fisheries Association was instrumental in creating the Southern Shrimp Alliance (SSA) to address the import controversy. The group consists of representatives from eight states in the south and southwest, in addition to two from the Vietnamese shrimping community.

Pressure from fishermen in the Southeast resulted in the implementation of two major emergency-assistance programs in 2003. The first was directed toward shrimpers. The federal government established a program of 35 million dollars for the southeastern and Gulf of Mexico states. North Carolina was awarded 4.8 million dollars to be distributed to shrimpers affected by the glut of imports. The amount of assistance given to each fisherman was based on landings information derived from per-trip data. One manager suggested that the average payout was 45 cents per pound, which was essentially a supplement to ex-vessel prices. Grace Kemp, of the NCDMF, informed me that 790 vessel owners qualified for this assistance and that the largest check was for $64,206, while the smallest was for just seven dollars. The average check was for about six thousand. A second emergency-assistance program, which I will discuss soon, was implemented for crab fishermen.

Foreign competition has obviously the changed the texture of the commercial man's work in regard to his economic returns from harvesting shrimp. We have also seen how the changing management structure has reshaped his work context and how governments, as in many other industries, have begun to develop permanent direct-assistance programs (also a matter I will discuss again in this chapter). But, historically, there is another piece in the industry's mosaic that the commercial fisherman has seen as a threat to his living. We have touched upon the subject in the previous chapter, but let us now explore it again in more detail.

EARNINGS AND ANIMOSITY AMONG FULL-TIME, PART-TIME, AND RECREATIONAL FISHERMEN When I first did field work on the shrimp fishery in the 1980s, I was struck by the intensity of the animosity between full- and part-time fishermen. I don't know why I should have been surprised because, by then, I had been consulting with the South Atlantic Fishery Management Council in Charleston, South Carolina, for several years. During that period I was asked to collect information from North Carolina to Florida on various types of fishing activities and had been exposed to a great deal of hostility among different fishing groups, including full- and part-timers. I guess the thing that struck me about North Carolina shrimpers was their feeling that they always were in a zero-sum game—the more others took, the less there was for them. This is not necessarily the case, but that is what was, and still is, perceived.

An examination of landings and value data over nearly four decades reveals that sharp drops in landings and value occurred in the 1974/75, 1977/78 and 1980/81 shrimping seasons. Another sharp decrease occurred between 1985 and 1990. Yet a study of data from 1967 to the present indicates that a steady *increase* in landings (and unadjusted value) has occurred, no doubt due to more efficient fishing techniques. Credit must be given to the management systems, as well, that were first put into place in the 1960s to protect the nurseries and to delay harvesting until shrimp reach an optimum marketable size. The year 2002 was a good capture year after the poor harvest of 2001, yielding 5.2 million pounds worth 11.9 million dollars ($2.26 per lb., unadjusted). By June 2002, nearly two million pounds had been sold at the docks, twice the norm for the immediate past years and four times the year before during the same period. Things began to even out as the year progressed, however, moving toward a 10-million-pound year and into the low 20 millions of dollars in value. And ex-vessel prices declined again.

Weather and contextual economic conditions clearly account for variations in stock abundance and economic returns. But this is not the way the fishermen have seen it. In addition to imports, full-timers blame part-timers and, to a degree, recreational fishermen for exacerbating the harvest and economic circumstances, even though climatological factors are recognized. When an estuary is opened for shrimping (dates are announced by NCDMF proclamations), hundreds of part-timers and recreational fishermen flock to the area along with the full-timers (see photo). The former come to the site with small boats and makeshift trawls. They are normally "try-nets," which are the test nets for larger vessels with the two- and four-barrel configurations that I will discuss more fully below. That is, the big boats put out a small, sixteen-foot try-net to see if the area they are in will yield a sufficient catch for the day's labor.

Full-time boat captains believe that individuals whose primary source of income is from other sources deprive full-timers of their inherent right to earn a living on the water. Part-timers and recreational fishermen complain

Chapter Six

An Opening Day for Shrimping and the Competition Begins. This aerial view of an opening day for inside trawling illustrates the various sizes of boats, some captained by part-timers, and the resulting density creates a great deal of hostility among the user groups.

of overfishing and blame the full-timers for declines in stock abundance. The fishing season affects competition among these three user groups. Historically, several hundred part-time fishermen have followed the opening of productive fishing grounds along the coast and competed with the full-timers who fish inside of the inlets. Fishing outside of the inlets has been prosecuted only by full-time commercial fishermen. The competition in the estuaries occurs mainly in July, when shrimping is at its peak.

Part-time and recreational captains have viewed the ownership of water resources to be as much theirs as anyone's. More than 80% of them are born in North Carolina, and more than half were raised in the coastal region. One out of ten has moved to the state and then retired here. Some have retired from non-military government work, while others have retired from the nearby military bases. They have fished part time and recreationally for enjoyment and supplemental income. Full-timers see fiberglass boats trailered by what they view as expensive cars and vans to "their" fishing sites with commercial gear aboard, and they become angry.

The Commercial Man

During my fieldwork in the eighties, there were more than 25,000 licenses issued that permitted the use of commercial gear. A little more than 4,000 of these were issued to full-time commercial fishermen (defined as earning more than half of their incomes from fishing). During this same period, between five and nearly eight thousand commercial part-time licenses, and between ten and fourteen thousand recreational licenses were issued, as we saw in the last chapter. This means that fishermen who defined themselves as "recreational" could buy a license to use commercial gear. By 1994, the largest decreases in licenses were among the part-time and recreational fishermen, while the number of full-time commercial licenses had increased by 50%. There is no doubt that the impending changes of the 1997 Reform Act prompted this shift. With the moratorium, those who felt a need to protect their rights to use commercial gear, and could qualify for the newly defined commercial license and ETS crammed the license bureaus.

It seems that the nature of the competition among the user groups was heightened among the full-timers, themselves, while it remained the same for the other two. And, as we discussed in the previous chapter, fishery managers and the legislature addressed the problem in all of the fisheries with rulemaking that changed the texture of the entire fishing culture, both in terms of who would be able to use various kinds of commercial gear and how they could dispose of their catches. As fishermen relinquish their licenses over time, other applicants who are on a waiting list can be selected to purchase the newly available licenses. Alternatively, they can purchase licenses in the open market or fish under a license owned by someone else.

A recent study of 259 Core Sound fishermen, two-thirds of whom are shrimpers, indicated that animosity directed by full-timers toward recreational fishermen was not widespread in that region (Cheuvront, July 2002:22–23).[4] However, Cheuvront's research involving shrimpers from Carteret County and southward revealed that some negative incidents with recreational anglers occur during the fishing year.

Chapter Six

In a recent survey of RCGL license holders, about 80% of the licensees indicated they did not feel they were in conflict with full-time commercials. Ninety percent of them felt the same about other recreational fishermen (Wilson, 2001 and 2003). Data from each of these studies may be an indication that the Reform Act is reducing, at least to some degree, the perceived direct competition between the groups.

HOW THEY ADAPT My fieldwork in the 1980s indicated that full-time captains were committed to fishing for nearly all (95%) of their incomes. At the turn of the twenty-first century, Brian Cheuvront (2002:19) found that Core Sound fishermen relied substantially less on fishing for their incomes. Just under half were totally reliant (48%), while slightly fewer than two-thirds relied on fishing for half or more of their individual incomes. His study of shrimpers in other areas of the state indicates a nearly identical pattern (Cheuvront, 2003). How these fishermen have adjusted to stock and price fluctuations is considerably different from that of their part-time and recreational counterparts. First, however, let us take a look at what, in fact, fishermen have preferred to do when landings appear to be normal or above.

ANNUAL-ROUND PROFILES Annual rounds—that is, the change of types of *fishing activities*—in particular, allow the fishermen to adjust to the natural cycles of marine resources. The rotation of the full-timer's activities during the calendar year, which I will now describe, has changed for some since I began studying the fishery. January and February are inactive for shrimpers, specifically, and fishermen, generally. Activity begins to pick up in April and peaks from May through October. Cheuvront (2002:14; and 2003) found a similar pattern among commercial fishermen throughout the coast in 2002. Chris Wilson (2003) found this to hold up among those holding RCGL licenses, as well, in 2002. July and August were the two months in which these fishermen were most active with shrimp trawls.

The *intensity* of shrimp fishing is as impressive as the *extent* of such activity, particularly among full-timers. From June through September, more than half of the commercial fishermen I studied concentrated exclusively on shrimping for a living. Another 10–20% were found to engage in one other activity (mostly crab trawling for small- to medium-size boats), but varies not only by coastal region, but also by exactly where the fishing occurs within a given region. Except for the month of July, when shrimping can be at its best, I found that shrimping and one other fishing activity accounted for all but a small portion of fishing effort of the commercial shrimpers statewide. Cheuvront found that about one-third of his sample of Core Sound shrimpers engages in clamming as their major second fishing activity.

Elsewhere, fishermen were changing their annual rotations to put more emphasis on fishing for hard crabs, especially in the Albemarle Sound. The lure of big money from crabs was as much an incentive to change as the lack of it was in the shrimp fishery, particularly for fishermen on the northern coast. This has been their response to the devastating impacts of shrimp imports. This also triggered an influx of Vietnamese immigrants into the Albemarle region who directed their activities toward crab fishing. The production and price results are eerily familiar. Beginning in 1978, the first disastrous shrimp year in the second half of the twentieth century, hard-crab production exploded. Landings rose from modest levels of slightly over 10 million pounds (unprocessed) to more than 34 million pounds in 1980, nearly 37 million pounds in 1990, and to an astounding 65 million pounds in 1996. Since then production has dropped off precipitously, dipping as low as 34.4 million pounds in 2002. Fishery managers are not sure of the reason for this precipitous drop. The aftermath of hurricanes in the late nineties may be related, according to respected scientists (Paerl et al., in Maiolo, et al., 2001:255–265).

The unadjusted per-pound price rose from 10 to 83 cents per pound. An increase in real value was realized during the nineties, from about 23 to 49 cents per pound. And that is a good thing. During the same period, however, imports of processed, packaged, and frozen crabmeat rose from 23 million pounds and 74 million dollars to 206 million pounds and nearly 880 million dollars. But the value of imports experienced a real decline. During the same period, adjusted per-pound values fell from $3.20 to $2.68. This may be an ominous sign that foreign competitors will continue to put pressure on the domestic market by increasing production and adjusting prices downward. One fishery manager I talked with in early 2003 noted that dealers had mentioned to her that the drop in North Carolina crab production occurred at about the same time the market had become saturated. That is, the dealers noted if they had received any more crabs at the dock, they would not have been able to sell and distribute them due to lack of market demand. As was the case with the shrimp industry, a second federal government emergency-assistance program was put into place in 2003. The state of North Carolina was awarded a total of nearly $1.8 million (out of a national total of five million) for cash contributions to crab fishermen, fish houses (dealers), and processors, (minus $92,000 for program administration). To be eligible for more than a $50 grant, recipients had to have been in the fishery during the past three years, and had to have landed or purchased 10,000 pounds of hard crabs within that period. Though not necessarily intended to exclude them, this program would not offer significant benefits for the shrimpers who fished for crabs on a small scale. On the other hand, they would have received supplemental assistance for shrimp landings. The smallest crab assistance check was for $50 (to a small-scale crab fisherman), and the largest was for $92,741 (to a large-scale processor). For all crabbers, the average check was for $237; for dealers, $1,256; and for processors, $24, 335.

During the period of the year when fishing activity is at its lowest, the Commercial Man has traditionally used his time to maintain and repair his boats and equipment. The effect of prices on his income can be seen in the changes in work activity during the off-season. Cheuvront found that one-third of the Core Sound shrimpers in his sample received income from non-fishing activities. These ranged from plumbing, carpentry, and general construction to driving trucks and working in other marine-related occupations (e.g., crewing on tugboats). The range of non-fishing activities was even greater in his study of shrimpers extending to Brunswick County (2003). Thus, we are seeing an expansion of the annual rounds to include increasingly more *non-marine* related activities as income supplements.

I found that, historically, *part-time* commercial fishermen, when such a formal category existed, concentrated less on shrimping (at least for sale) than their full-time counterparts. In the peak month of July, about one-third fished exclusively for shrimp. Shrimping, targeting finfish with gillnets, and crabbing represented the most popular activities. From July to October, all three were frequently combined. And while a very few fishermen did not own a boat; over 90% were boat owners.

During the period when the Reform Act was under consideration, the issue of license costs came up which sheds some light on the importance of commercial fishing for the part-timers who intended to move up to the SCFL, priced at $200. However, a vessel-registration fee is also required for each boat. This fee is based on the size of the boat. So, a boat of over eighteen feet and up to thirty-eight feet is charged $1.50 per foot. Thus, an owner of two vessels intended for commercial use, averaging say, 25 feet (x 2) would pay 50 x $1.50 or $75. Vessels of more than thirty-eight feet and up to fifty feet are charged $3 per foot, and those of fifty feet or more are charged $6 per foot. At some of the public hearings on the Reform Act, the part-timers complained that the large-vessel operators were being given an unfair advantage in the price schedules because small-vessel operators, as most part-timers are, had to buy two or more licenses, while the large operators only had to buy one. Some full-timers may not welcome the comparison, but apart from the fact that the cost structure does not seem to be out of line, such testimony is further evidence that many part-timers are serious commercial fishermen and had competed heavily in the past with the full-timers. Furthermore, their style of fishing could qualify them to transition into the SCFL category from the part-time classification under the old scheme, even though fishing is not their main occupation. I examined this issue in the prior chapter and do so again below. It is an issue that constantly surfaces and permeates the industry.

Recently acquired data indicate that some changes in activity have occurred among those who used to be classified as part-time commercials—in actuality recreational fishermen who occasionally sold portions of their catches. Those fishermen holding the newly devised recreational license allowing the use of commercial gear (RCGL) and who own a shrimp trawl constitute just under one-third of the population. They do mirror their full-time counterparts in when, during the year, they use shrimp nets most frequently. However, the number of them who shrimp during the height of the season appears to have dropped to one-fifth of the license holders. About one-third now trawl for crabs, and most of the remaining trawl for finfish. These are the ones who, for one reason or another (not the least of which is their inability to qualify for the SCFL), are now grouped with true recreational fishermen.

GEAR TYPES AND ACTIVITY LEVELS While the overall trend toward larger vessels discussed in Chapter Four seems to have subsided, some ocean trawlers extend more than ninety feet in length. Improvements in gear have continued to find their way into the fishery. Navigation for fishing has been enhanced by the Global Positioning System (GPS). Nylon nets have gradually, and now entirely, replaced the traditional cotton gear. However, as I noted in Chapter Two, the most important technological changes since the mid-fifties have been the adoption of double rigs, which began in earnest in the 1980s. First developed in Texas in 1955, the *double* rig consists simply of putting out two trawls instead of one, thereby increasing the harvesting capacity at pretty much the same cost (one boat, same crew size, and same time on the water). Fuel costs do increase, however, because of the added drag. The Morehead City Shipbuilding Corporation may have been the first North Carolina company to outfit larger vessels with double rigs, launching one in 1957.

Twin trawls were introduced to the region in the 1970s. Two trawls are attached to a single pair of doors, which spread the mouth of the net. The trawls are then joined at the head- and footropes to a neutral door, which is connected to a third bridle leg. Such a rig is claimed to increase efficiency up to 25% over the double rig, is lighter in weight, easier to handle, can be towed with less speed, and can make sharp turns without tangling (SAFMC, 1981:8–60). But twin trawls are not necessarily larger than a double- or even single-rig configuration. For example, a twin trawl may consist of two thirty-five-foot nets, whereas a single may use a seventy-footer. So called "four-barrel" rigs put out attached twin trawls on each side of the vessel.

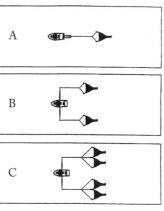

Net configurations typically used in the shrimp trawl fisheries. (A) single conical net, (B) double set, and (C) twin set. Adapted from ASFMC 1996.

Mr. Billy Burbank of Florida developed the "three-winger" or "mongoose" in the late 1970s. This device has a triangular tongue, or wing, "attached along the midsection to a center towing cable" (SAFMC, 1981:8–60). Whereas conventional trawls spread from 50 to 75%, the mongoose spreads to 80% or more. Mongoose trawls have a greater height, as well, resulting in "greater catches of high moving white shrimp" (SAFMC, 1981:8–60). There appear to be savings associated with the use of smaller doors and decreased engine speed.

Historically, there have been three distinct patterns of shrimp fishing with trawls in North Carolina. Each is associated with boat size, gear configuration, location, and the nature of the annual round. The first is the small-craft fishermen using boats up to twenty feet. I found that about 13% of those fishermen are full-timers and that, in addition to shrimping, about half of them dredge for scallops when prices and stocks are favorable. One-third of them harvest crabs with wire-mesh traps, and about another third, until recently, were involved with channel-netting. Two-thirds set gill nets for bluefish, croaker, and trout. Just under half were found to gig for flounder at night, a technique requiring the fisherman to walk or pole his boat along the shoreline, find a flounder lying on the bottom, and then spear the fish with a three-pronged device at the end of the pole.

Statewide, the majority of full-time commercial fishermen (about two-thirds) own boats of nineteen to forty-eight feet, a number similar to that found in the Core Sound study. Those who shrimp with those boats do so inside the inlets. Boats over thirty feet have used the double rig since the 1970s. For all fishermen who use trawls in the Core Sound for most of their income, fishing for crabs has been the second most lucrative activity. Those who specifically target shrimp as their main source of activity and income, however, do not use their trawls for other species, except in rare cases (Cheuvront, personal communication with the author, April 2003).

The vessels over fifty feet, owned by about a quarter of the commercial fishermen, fish outside the inlets. They are equipped with all of the modern electronic devices—VHF radios, GPS equipment, and sophisticated depth finders—and some have radar and/or autopilots. Unfortunately, I found that very few were equipped with survival suits.

The larger vessels can travel long distances to fishing grounds as far north as Georges Banks (off Massachusetts) and south to Florida. Many shrimp along the South Atlantic coast during the season. Double rigs have been popular, historically, but larger vessels also have been found to use twin and mongoose trawls. Finfishing and scallop and crab trawling are the only other activities in which these boats have engaged, with one exception I found some years ago. A few of them were experimenting with trap fishing on reefs, but federal regulations put a stop to that practice.

Before the Reform Act of 1997, part-timers could sell their catches. Some of these qualified to get an SCFL. Others did not, as we saw in the last chapter. The latter were able to obtain, however, an RCGL, as I have noted.

Historically, all of the part-time fishermen have fished with boats of forty-eight feet and under. The larger fleet in this class consisted of mostly local and, even then, former full-time fishermen. About two-thirds of this class of vessel were in the eighteen-foot range. And at any given time, about two-thirds were inactive—that is, until the shrimping season heated up. A very few of these did use double rigs and twin trawls, which drove the full-timers up the wall. Most used single trawls of the try-net variety. Until recently, clamming was the second most popular type of fishing, followed by gillnetting, oystering, then crabbing, and some scallop dredging, when it was allowed. Flounder gigging was, and still is, a very popular activity among the small-boat fishermen during the summer months.

An examination of the location of the total (not just the active) RCGL licenses in 2001 indicated that about half the licensees were in either coastal or near coastal counties with rich fishing traditions. Brunswick, Carteret, and New Hanover led the coastal counties, while Craven led the near-coastal counties, followed by Columbus and Beaufort. The number of years of experience of these license holders is a good indication of the impact of the Reform Act in shifting a number of part-timers to the recreational category. About 60% of them had more than ten years of experience with commercial gear, and just over one-third had twenty or more years of experience. Most of the respondents studied by Chris Wilson (2003) have spent their entire lives in North Carolina. Of them, 96% were found to be male and 93% white. About three out of four of these license owners have total household incomes of more than $30,000 per year from jobs (other than fishing, by definition); over half have annual household incomes of more than $50,000. Using an inflation index, there has been no change here, either, from the data I gathered in 1980. The comparable numbers for commercial fishermen show that only 13% of their households had total annual incomes of more than $50,000 and that

60% earned *less* than $30,000 annually. Sixty percent of the commercial fishermen earned less than $15,000 from fishing activity only (Cheuvront, 2002)! That is an astounding number that deserves further examination.

Based on the data at hand, I would venture a guess that about two thousand fishermen who would have been classified as part-timers before 1997 are now SCFL holders. It appears that these are fishermen who continue to move in and out of shrimping on an opportunistic basis, just as they had prior to the adoption of the new system. Thus, a major source of competition and conflict is still present. Included in this category are those who would rather fish full time, but cannot make the kind of living they want if all they did was fish. This notion is consistent with other research on commercial fishermen by Griffith (1996b) and Garrity-Blake (1996). Some move between fishing and non-fishing income sources but maintain an identity with fishing as a way of life. This can be seen as an extension/modification of the traditional annual-round profiles.

It is more difficult to estimate the number who shifted into the RCGL category, but I would suggest that the remainder of the historic five to six thousand part-timers (about three thousand) applied for the newly created RCGL. Most of those who would have paid a modest fee to "possibly" use commercial gear in the past now must purchase a limited license for capturing shellfish and use gear, for example, like a crab pot.

What is uncertain under the current regulations is how much seafood caught by RCGL holders is illegally sold. It happens, make no mistake about it. I have always believed that seafood catches and sales, in general, and shrimp landings, in particular, are underreported by as much as 25%. Clearly one of the reasons that the laws against this illegal practice go unenforced is that the jurisdiction over it is not well defined. Also, there are not enough marine law enforcers to go around as it is, let alone take on a task like this one. And, finally, there are *many* people involved in the underground sale of seafood, which has the effect of depressing any interest in dealing with it. There is somewhat of a "no harm, no foul" mentality. What this means to me is that more damage might be done to the fishing culture by dealing with the issue than by simply ignoring it, and I find it hard to disagree with that notion. It is a very tough way of life out there as it is. We probably don't need to make it any more so. The full-timers trying to scrape out an existence in increasingly difficult circumstances may not agree, however.

BOAT CREWS John Bort and I found out a long time ago that there is no such thing as a "typical" crew or crewmember on North Carolina shrimp boats. The fishermen work alone on trawler skiffs and channel-netting vessels. Medium-sized trawlers have crews of two or three, while the larger boats have four. More than half of the crewmembers I studied also have their own boats (averaging twenty-one feet). Of these, one-third shrimp on their own when they are not working on the larger boats. Clamming, flounder gigging, and gill netting are also popular activities.

Most of the medium to large boats in the shrimp fishery have crewmembers who perform a variety of tasks before, during, and after a trip. These include changing the rigging, repairing equipment, cooking, setting and hauling nets while fishing, culling the catches, and heading and icing the shrimp (on the large vessels). Crewmembers also participate in decisions involving the operation of the boat. Their inputs vary according to the size of the

Commercial Man Mending His Nets.

vessel. The need for crew coordination and the risks associated with the work tend to create more democratic patterns than in most other occupations, according to another colleague, Richard Pollnac (1982:228).

Nevertheless, directing the activities of the larger boats is more typically in the hands of their captains. In cases where boats are owned by dealerships or partnerships, the captains and other owners usually decide the type of fishing they will do, the areas to be fished, and other matters. The captain decides specific daily activities within that framework, albeit with some input from the crew.

Payment systems vary somewhat, but are usually based on a "share system." Brian Cheuvront (2002:16) found that Core Sound fishermen on vessels with at least two crewmembers use a share system in about 71% of the cases he studied. John Bort and I found the most typical share system to look like this: After the sale of the catch at the dock, fuel and grocery costs are deducted from the total. We found that 5% of the remainder went to the captain before any other allocation occurred in about half the cases. Between 25 and 50% of the money left, called "dead shares," goes to the boat and equipment (owner); and then the captain and the crew share in the remainder equally. In cases where the captain does not get 5% off the top, he gets between 1.5 and 2.0 shares. The crew usually receives "shack money" from the trip. This is the revenue earned from the sale of the incidental catch. Some years ago, shack money in the deep-sea scallop fishery referred to cash received from the sale of undersized product, which was hidden in the boat's sleeping quarters.

Commercial Man Unloading His Catch of Shrimp.

Let's walk through an example. Assume that sales from a trip yielded $10,000. Under the system we described, let us further assume that $1,000 was spent on fuel and groceries. Four hundred and fifty dollars would be set aside for the captain, and $8,550 would remain. Of that, $4,275 would go to the owner(s) (the dealer and/or the captain), and then the captain and the crew (let's assume the total is three) would each receive $1,425. If the captain owns the boat, then he would receive $450 (5% off the top) + $4275 (ownership) + $1425 (share), for a total of $6,125.

Under the new licensing system, someone who crews long enough to establish that his/her main source of income is from fishing can use that information to gain access to the queue to get his/her her own SCFL. Previously, that person could purchase any one of the three licenses that were available—full time, part time, or recreational—that permitted the use of commercial gear and, for the first two types, the sale of product.

EXPECTED EARNINGS AND LEVEL OF FISHING ACTIVITY Early in our research, John Bort and I found that expected earnings significantly affected the fishing activities of both full- and part-time fisherman, but in different ways. Under the more generous, historic licensing system, which allowed part-timers to use commercial gear and sell their catches, their activity would increase or decrease in direct proportion to their expected earnings. Three-quarters of them earned income from regular, steady jobs outside of fishing. The remainder moved in and out of other jobs as opportunities developed or disappeared. Their annual incomes outside of fishing averaged more than $18,000 in the early eighties, and had nearly doubled by the time the Reform Act was put into place. Remember that more than a few of these were federal and state jobs, which also offered good benefits. Thus, while part-timers were earning extra income from fishing, a bad harvest year was cushioned each month by that steady paycheck from their regular jobs.

We found that fishing *effort* among full-time captains remained steady, regardless of expected earnings. Bort and I also explored other issues—such as who owned the boat, the number of partners, if any, and whether the fishermen were crewmembers of other boats—to determine what impact they might have on activity. None was found to make a difference.

Poor shrimp harvests, however, eventually did change the *types* of fishing and other activities in which the Commercial Man was involved. We first examined data from three specific "down" years—1978, 1979, and 1981—to get a fix on this matter. After the 1981 season, while some of the full-timers quit fishing because of health or retirement, about 5%, tired of the uncertainty, took land-based jobs. During that same period, one-quarter of the part-timers indicated that they had quit fishing permanently. "Shoot, I can buy the damn shrimp a whole lot cheaper than ketchin' 'em," one told me. Recall that in that timeframe the part-timer could exit and then return without any difficulty. Today, if he had an SCFL and were to give it up, he may not get it back. According to others we interviewed, most had quit because of the condition of the seafood stocks and markets. The return for their efforts, plus the perceived prohibitive cost of continuing (especially fuel), provided the incentive to look for other things to do with their spare time.

I was able to retrieve data on two grade school graduating classes from Harkers Island that shed some light on the subject of the effect of poor harvests among potential new entrants into commercial fishing.[5] Harkers Island is a community in which fishing and boat building dominated the economy for nearly a century. Of the twenty-five 1978 graduates, fifteen of whom were male, only four went on to become fishermen. Another became a dredge-boat operator. By 1998, the number of graduates had dropped to sixteen, and I could not find one who had become a commercial fisherman. Cheuvront asked one of his samples of shrimpers how they would respond to inquiries from young men and women about commercial fishing as a way to make a living. His data indicated very little likelihood of a positive recommendation.

Among the full-timers I interviewed, nearly half chose the path of altering their annual rounds beginning in 1979. Exhibiting, once again, what I call the "ethic of opportunism," some of the captains of larger boats became heavily involved in the New England sea-scallop fishery. Their choice was reinforced by the successful harvest that year and two years later, when the North Carolinians experienced a great deal of success scalloping on the Georges Banks. Some fished in New Jersey, as well. Large steel-hulled dredgers from Manteo had entered the fishery before the new round of shrimper entries. The shrimpers who entered the fishery actually used their shrimp nets to harvest scallops, but they fished shallower water than where the dredgers could work. The term they use for this activity is "shell-stocking." Stock, in the shell, would be brought on board; then the boats would steam back to North Carolina and offload in one of several locations for shucking. Dredgers use their crews to shuck on board while the gear goes back in the water for more harvesting. It was during this time that illegally "shacking" undersized scallops became a standard practice among dredging crews.

It wasn't long before the New England fishermen complained that the North Carolinians were damaging the scallop stock by recruitment overfishing. A great deal of controversy occurred among managers, scientists, and those in the fishing industry, and I was right in the middle of it. Some of those involved wanted to restrict fishing for scallops to those who "historically" had been in the fishery. The scientists working in New England argued that limits on size and severe tolerance restrictions needed to be placed on harvesting. I was commissioned by the South Atlantic Fishery Management Council to determine the impact of such limits on North Carolinians. I was able to show that the proposed limits put the New England dredgers at a competitive advantage because the dredgers could fish in the deeper water while the shrimpers could not. This would effectively shut the North Carolinians out of the fishery. I was also able to show that the proposed restrictions would result in the loss of hundreds of shore-based shucking jobs in North Carolina. I wrote a report on these matters, based on research funded by the North Carolina Sea Grant Program (Fall, 1981).

We lost the battle. Shortly thereafter, continuing worries about possible stock declines resulted in the adoption of federal regulations governing size limits, which redirected scalloping efforts from New England to the expanding calico-scallop fishery in Northeast Florida (see Maiolo, Fall 1981). By 1981, fifty vessels, most from Carteret County, were fishing there. Twenty-six of these had made the conversion from the north to the south.

Other captains fished in South Carolina. Johnson and Orbach (1990) published results from their wonderful study of that aspect of the migration. Their study found that most of the boats in their sample were medium-sized trawlers of forty-five feet or less. They also found that migratory fishing patterns are influenced not only by stock fluctuations, but also by friendship connections and fishing "networks."

Captains from the Southport area continued to seek shrimp, but did so in South Carolina and in the distant waters of the Gulf of Mexico. Those who entered the Florida scallop fishery revised the shell-stocking concept by offloading right at the docks in St. Augustine or Cape Canaveral rather than returning to North Carolina to do so. There the catch was loaded into semi-trailers, which then headed for North Carolina coastal communities where fishermen's wives would do the shucking. The truck drivers would stop over briefly in Georgetown, South Carolina, to restock their trailers with ice to keep the shellfish fresh.

An enterprising North Carolinian, along with local Florida dealers, opened up several very efficient scallop-processing plants, one of which could machine-shuck scallops at the rate of six hundred gallons an hour. In North Carolina, it took practically a whole day to shuck that amount, even with machine processing!

It is important to recognize that a lot of full-time fishermen have severe limitations in their land-based occupational alternatives. Some of these individuals just simply cannot reorient themselves away from their work on the water. They see non-fishing work as a "job" as compared to their view of fishing as a natural calling stitched into their coastal way of life. Many know nothing else, and they don't want to know anything else. You fish or get a job (see Garrity-Blake, 1996, for further discussion of this topic). Interestingly, during the past few decades, increasing numbers of fishermen's wives have entered the labor force in "regular" jobs, as opposed to seasonal work in seafood processing. Garrity-Blake found that this trend toward a steadier source of spousal income allows some commercial fishermen to stay on the water. Indeed, some women indicated they wanted their husbands to stay on the water, if only to keep them happy.

Other fishermen hate the prospect of working for someone else or in larger work groups of people who are not from their own communities. Working in another region is okay on a temporary basis, so long as the work is fishing or something close to it, as Johnson and Orbach discovered in their study and Blomo, Orbach, and I found out in our research on the menhaden industry (1988:41-60). But migratory fishing puts a tremendous strain on family relations.

For those who feel compelled to leave commercial fishing to pursue other income-producing activities, Garrity-Blake (1996:2) revealed that higher paying, white-collar employment is generally not an option due to age and educational issues, as well as economic uncertainties. Those who do leave commercial fishing, including those who continue to fish opportunistically, look for blue-collar employment in trucking, auto mechanics, construction, and government (e.g., the Port Authority in Morehead or Cherry Point in Havelock).

Most of the shrimpers who stay with fishing will move in and out of other alternative fishing styles, depending upon the equipment and experience they have at their disposal. Community networks, mainly based on kinship and friendship, play an important role here. A great deal of experimentation with clam-kicking began after the 1978 season, prompted largely by the increase in the price of clams due to the devastation of the crop in the Northeast by weather and pollution. Others tried oystering, and scallop dredging in the estuaries. Scalloping was unsuccessful, however, because of the glut of product coming from, you guessed it, Georges Banks and Florida. In Florida alone, the harvest of processed (shucked) calico scallops rose from 4 to 20 million pounds in just three years.

Chapter Six

Borrowing money is anathema to most North Carolina fishermen, something to be avoided if at all possible. When borrowing is necessary, other than for a home mortgage, full-timers borrow small amounts from dealers, friends, or relatives in order to buy equipment or make repairs. Most such loans are from dealers and are interest free. However, the fisherman will then allocate 10% of his income from future catches toward the debt until it is retired. In such instances, the Commercial Man is expected to sell his future catches exclusively to the dealer who made the loan, at least until the debt is repaid. We will explore this further in the next chapter.

With a couple of exceptions, very few fishermen other than those in the menhaden industry have participated in government assistance programs. Welfare in particular is to be avoided like the proverbial plague, even though some fish dealers dispute this, as we shall see in the next chapter. Normally, those who take advantage of specialized loan-assistance programs are large-boat operators, many of whom are both dealers and boat owners.

But other programs have been in place for decades to provide boat loans for purchases and repairs and to provide tax shelters, which produce substantial savings.[6] I have found very few small-boat operators who

Refurbishing An Aging Shrimper.

participate in any of these. In fact, I have never found one fisherman in the state who ever took advantage of any of the tax-shelter programs. And, even more surprisingly, I found that the accountants for some of the largest dealers in the state were not even aware that such programs existed.

With regard to cash grants, many people in the state's commercial fishing industry who were directly impacted by the devastating Hurricane Floyd in 1999 received more than $7.5 million dollars in aid from the State of North Carolina (Street and Williams, May 2001). Following the disaster, the General Assembly and Governor Jim Hunt committed more than $836 million in state funds to the storm's victims throughout the eastern part of the state, and those in the fishing industry were not left out. It took a couple of years, but by 2002, assistance checks were sent to nearly a thousand fishermen, dealers, and processors. These checks ranged from a few hundred to over five hundred thousand dollars (to the state's largest menhaden-processing company, which, in turn, shared the proceeds with its employees). The monetary disbursements were based on dependency and covered lost income and gear. The typical grant was in the several-thousand-dollar range (NCDMF, March 2001).

The fact that this program, and those I described earlier in the chapter involving the shrimp and crab fisheries, provided help to recover from a natural disaster and was offered with few strings attached, surely made such assistance more readily accepted by many. However, and not to my surprise, a number of fishermen *refused* the help. In the case of the Hurricane Floyd relief, many simply did not apply. Grace Kemp, of the NCDMF, informed me that sixty-seven fishermen declined to participate in the shrimp relief program; their typical reason was that they did not want "a handout." They had gotten along without the government before. There was no reason to change now. At the time of this writing, the number of refusals in the crab assistance program was unknown, but they were coming in. In some cases, the refusals were probably due to some, and perhaps justifiable, concern over what government people might find when looking at landings and

Chapter Six

revenue data. But the pattern of refusing "a handout" fits the values I have seen in the fishing community since I began this research journey. The fishing community is its own main source of support for the hunters and gatherers of the sea; they have little use for government bureaucrats who have never set foot on a commercial fishing boat.

In general, the North Carolina commercial fishing fleet, with the exception of the larger steel-hulled trawlers, and the shrimp fleet in particular, shows the effects of poor revenues. And this puts the state's full-timers at a further competitive disadvantage with fishermen from other states. During an extensive field trip I took to the central coast in the winter of 2003, I was able to find fewer than a half-dozen medium-sized shrimp trawlers that looked like they had been built within the previous ten years.

Varnamtown Fleet in 2003.

RESPONSES TO RESOURCE MANAGEMENT Historically, fishermen have had a difficult time accepting regulation of their activities. After all, one of the reasons they choose fishing as an occupation is because they embrace what they see is freedom to explore and harvest. The one limitation they can accept is based on what nature can provide. They have had trouble understanding the fishery biologists' and economists' models and jargon—MSY (Maximum Sustainable Yield), OY (Optimum Yield), recruitment and growth overfishing, internal rates of return, discounted value, critical size, and so on. They have their own set of explanations about why fish behave as they do. This is something of a folk science, which some of my colleagues have christened *ethno-ichthyology* (fishery science of the folk or ordinary people). Garrity-Blake reported that the most common reason her sample of former commercial fishermen gave for seeking alternative sources of employment was "bad fisheries management" that ignored their input (Garrity-Blake, 1996:2). In spite of this, many members of the various user groups see state jurisdiction over water resources as a necessity. Much of this support, unfortunately, has a self-serving basis.

Compared to part-timers and recreational fishermen, not very many full-time fishermen actually participate formally in the management process, in spite of the NCDMF's efforts to include them. I have been present when large numbers of shouting fishermen have attended public meetings. But, in actuality, fewer than one-third of full-time fishermen have ever attended one NCDMF meeting. When I attended federally sponsored meetings in North Carolina on the sea-scallop issue, only five fishermen attended each hearing. Other federally sponsored hearings in North Carolina that I have attended averaged ten fishermen, at most. Hundreds of fishermen and shuckers earned a large part of their incomes from scalloping, and the proposed regulations threatened to, and eventually did, exclude their participation in the fishery. This pattern of low attendance occurred in spite of the fact that every fisherman and processor was contacted and urged to attend the meetings. If they had not been, their participation might have been zero. I know that a great deal of planning goes into scheduling meetings in an attempt to accommodate the working needs of fishermen.

Chapter Six

Just as their federal and regional counterparts do, the state's fishery managers try to schedule meetings at times and places to accommodate the fishermen. Those who do attend the hearings, mostly large-scale operators, tend to be opinion leaders in their respective communities. They can be quite vocal and intimidating to their small-operator counterparts. However, some of the smaller operators will occasionally take a more confrontational approach. I attended several meetings about clam-kicking, clam-potting, and long-hauling (in Pamlico Sound) and wondered if violence was about to break out. These meetings involved cases in which groups of full-time fishermen, divided by differences in what species they were targeting and what kind of gear they were using, *accused each other* of interference. For example, those who harvested with crab pots complained that the long-haulers (those using twin boats hauling large nets to capture finfish) disrupted their work. Long-haulers argued that the opposite was the case. Some meetings I attended before the Reform Act went into effect drew a large number of part-timers who were often mistakenly perceived by other attendees, including the press, to be full-timers. The best attended meetings were those on shrimp, crab, and clam-kicking harvesting methods and seasons. The deep-running divisions among the various user groups were very apparent.

I often thought that an almost fatalistic ethic seems to prevail among many full-time commercial fishermen. With some exceptions that I know of, many do not feel comfortable in direct confrontation of any kind and are even more uncomfortable dealing with issues they see defined in complex, seemingly unintelligible language. One of the most sensitive fishery managers I have ever known—let's just call him Bruce—really understood this problem and did everything he could to make his presentations in language the fishermen could comprehend. Mike Orbach and I saw this issue developing right after the Magnuson Act was adopted nationally. In our 1982 book (1982:6) we noted:

> For fishing peoples, communities and industries ... modernization is not only the process of adopting new physical materials, objects

and organizational technologies in their occupation and lifestyles, but also the process of learning how to work within the new political and administrative constraints and requirements imposed by the upward aggregation of management regimes.

Historically, even the hearings on the most radical changes on access into fisheries, including the public meetings on the Reform Act, were not that well attended by full-time commercial fishermen, according to witnesses. Thus, while the North Carolina Commercial Man has frequently demonstrated a remarkable capacity to adapt to the changing physical and biological characteristics of his environment, his adaptation to the evolving economic, political, and administrative constraints may still be lagging behind. This is the case in other states as well.

Two notable exceptions are the Draggermen's Association in New England and the Organized Fishermen of Florida (OFF). I found these groups ready to respond to issues in a manner resembling the way labor unions have historically dealt with industrial management—through organization, discipline, tradeoffs, and most importantly, hiring skilled professional specialists to anticipate and deal with impending issues.

Sport-fishing groups have become a political force nationally, regionally, and locally. Fishery management plans, especially those dealing with habitat, and some of the more popular game fish, have been greatly influenced by organizations like the Florida League of Anglers (FLA). North Carolina's resource managers have been responsive to groups like the Raleigh Sports Club, the North Carolina Chapter of the Coastal Conservation Association, and the North Carolina Coastal Federation in managing finfish, shellfish, and, of course, shrimp through trawling regulations. This is very much related to the demographics of sport fishermen, especially in regard to their education, occupations, and urban lifestyles, which are significantly different from those of full- and part-time commercial fishermen. These attributes make them more inclined to participate in formally structured and purposive voluntary associations. Fishermen in small coastal littorals are, in general, not so inclined.

Chapter Six

In the past, the North Carolina Fisheries Association (NCFA), established in 1951, has had difficulty gaining enough traction to rally commercial fishermen into action. Small-scale operators believe that the large-boat owners manipulate the organization to their advantage. This may not, in fact, be the case, but that is what many fishermen have perceived in the past. Many small-boat operators also don't feel they can take the time off the water in order to devote it to any organization. Despite this, the association's meetings seem to be fairly popular. We found that 75% of the state's fishermen have attended at least one meeting but that, historically, only about 25% have been members of the NCFA.

One of the instances in which the NCFA exerted a great deal of influence was in its support of much of the Reform Act, which is seen as a major victory for the Commercial Man. While critical of some of the provisions of the Act, including language pertaining to some of the license provisions, the NCFA was supportive of the Moratorium Committee's work and clearly supported the moratorium on commercial licenses. Currently, the organization is actively involved with the Southern Shrimp Alliance in gathering data on shrimp imports. It is clear to me that their intent is to try to muster support to somehow reduce import levels.

The relationship between the fishermen and the people to whom they sell their catches is interesting. Both the Commercial Man and the dealers with whom he interacts almost everyday face issues that are not always perceived by the other. How this interaction plays out is the subject of our next chapter. After that, we will wrap up the book by locating the shrimp fishery's major players within the community context in which they operate.

The Carolina Cannonball[1]

We have seen that North Carolina fishermen do not fish exclusively for shrimp. There are those who concentrate on harvesting the resource, but they also engage in a variety of other activities that directly or indirectly impact their attempts to harvest shrimp. And recently, more are looking for ways entirely outside of fishing to supplement their incomes. They look for work in construction, sometimes with government agencies in the area (e.g., the Cherry Point Marine Corps Air Station in Havelock), as ferryboat operators, and so on. A few see this as a permanent change. Many want it to be temporary, just long enough to get by the hard times and get back to where they belong—fishing! More fishermen's wives are also in the workforce than I have ever seen before, and these are not just wives of the young fishermen. Some of my respondents who are in their sixties now send their wives off to work every day as clerks, secretaries, and waitresses. On occasion, but in ever decreasing numbers, wives in the fishing community work as shrimp headers and finfish processors and packers.

Those who buy seafood at the dock and process and/or sell it to others must be viewed in a similar light. The label "seafood dealer/processor" covers a wide range of people doing a lot of different things. Just like their commercial fishermen counterparts, some dealers operate only on an opportunistic seasonal basis, while others are open for business throughout the year. Some buy and sell only a few species, while others cover as wide a range of products as they can get their hands on. Some buy the product, pack it, and then ship it on to another company for processing or take it directly to market. Others buy, process, pack, freeze, and ship. A dealer may operate out of a small building at the back of his home lot, sell out of a truck, or own a large business occupying several buildings in which processing operations can change in a heartbeat as new products become available. And I have found a number of dealers who have, just like their fishermen counterparts, developed business ventures outside of fishing. But if the fisherman is considered to be the door to the fishery, the dealer, regardless of the size or complexity of his business, is the hinge that connects the door to its market frame. This chapter will examine the full range of marketing and processing activities that operate in the coastal region of North Carolina.

Shrimp is an important element in the marketing operations of the state's seafood industry and, to varying degrees, in the Southeast region. From about the time the otter trawl was first introduced in the early 1900s through 2002, North Carolina's contribution to the total value of the Southeast's shrimp harvest ranged from 8 to 36%, as noted in Chapter Four. The more likely modern range is between 15 and 25%. However, the contribution to the national total averages only 3–5%.

The contribution to the *state's* total fishery revenues has varied from as low as 5% before 1900 to as much as a record 42% in 1971. It became the state's most valuable fishery in the early 1950s, when it surpassed the capture of menhaden. Even though a number of record-value years have occurred during the 1990s and again in 2000, its recent contribution to the overall value of the state's fisheries has been averaging less than 20%. The major reason for this is the sudden increase in value of the blue-crab fishery. From a contribution of $1.3 million in the early seventies, that fishery has grown into an industry whose ex-vessel sales approaching $40 million are not unusual, and annual sales in the mid-thirty-million-dollar range are becoming a common occurrence. By 1994, the hard-crab fishery had become the state's most valuable. Nevertheless, shrimping remains a strong second and still involves almost all of the standard dealer/processors on the coast, if not in the state.

Differences among dealers are influenced by the availability of the fin and shellfish they sell and the size of their operations. The larger operations—some, but not all of which are located in the larger coastal communities—operate during the entire year. The medium- and smaller-sized ones, especially those in the more isolated coastal communities, do not. Larger operations have access to substantial quantities of locally harvested fin and shellfish, not just seasonal crops such as shrimp. Some of them changed their activities in the late eighties to include the sale and distribution of imports. The pre-1980s traditional patterns plus the inclusion of imports, allow processor/dealers an even, relatively predictable pattern of business activity throughout the year, even though the specific products and markets change during the annual cycle. I found that some medium-sized dealers had begun to follow this pattern in the eighties.

The larger operators have always engaged in more diverse handling activities than their smaller counterparts, who do little or no processing. The latter will head, sort by size, and pack shrimp for shipment to another dealer or directly to market. If the opportunity arises, they will pack and sometimes freeze finfish and then ship the product out to other dealers and processors. They also do the same with shellfish, such as clams, oysters and crabs, but do not normally deal in scallops, which go directly to a shucking house. But medium- and small-scale operators do participate in the crab fishery on a regular basis, acting as buying agents for the larger operations. This activity does not interfere with their handling shrimp at the same time.

While most of the larger operators process both fin and shellfish, others ordinarily do not, except on an opportunistic basis. They may differ also in the varieties of fish they handle or fillet. But they all pack large quantities of fish for distant markets and will either own the necessary shipping trucks or be serviced by the buyer's/distributor's large trucks or by a third-party trucking company. Beginning in the eighties, some of the

Chapter Seven

larger fish-processing houses began to export some species, especially flounder to Japan.

Data from 2001 reflecting the range of dealer activity indicated that 219, or about one-quarter of the licensed dealers in North Carolina, handle shrimp. One hundred and sixty companies in five counties—Brunswick, Carteret, Hyde, Onslow, and Pamlico—handled about three-quarters of the state's landings in terms of value. Most of the landings and the greatest value came in through Carteret County, followed by Pamlico and Hyde. They paid an average ex-vessel price per pound of $2.22.

Table 7.1 shows data and values for total amount of edible seafood and number of dealers in counties where more than a million pounds were handled. The "Pounds Landed" column indicates ex-vessel (i.e., offloaded) weight. The "Value" column shows the total amount paid to boat at the ex-vessel point of exchange.

Dare County leads with more than 31 million pounds and nearly $25 million in ex-vessel sales. This includes, in addition to normal offloading of fish from local waters, sales both by North Carolina fishermen with large vessels who offloaded catches in North Carolina that had been made in waters outside of the state and by out-of-state fishermen who sold their catches to North Carolina dealers on an opportunistic basis. Carteret and Hyde Counties follow with substantially more ex-vessel landings and dollars than other counties. Of interest, too, are the high numbers of dealers in the coastal counties of Dare, Carteret, Brunswick, and New Hanover. This is explained by the fact that these counties are the locations of the smaller roadside dealers who orient their businesses toward travelers, as well as home to merchants whose main focus is something else but hold many dealer licenses in order to supplement their revenues by selling seafood. Finally, note that New Hanover County rests near the bottom of the table, both in terms of landings and values. New Brunswick is eighth in landings and fifth in value. When we get to Chapter Eight the significance of these positions will become more evident.

TABLE 7.1* Ex-vessel Landings and Values of Edible Fish per North Carolina County: 2001

County	No. of Dealers	Pounds Landed	Value
Dare	90	31,208,627	$24,975,642
Carteret	125	9,825,893	$14,102,023
Hyde	42	9,063,900	$8,181,350
Pamlico	50	4,948,362	$6,460,528
Beaufort	43	4,771,826	$4,902,486
Tyrrell	11	2,944,281	$2,464,712
Onslow	48	2,773,212	$5,457,258
Brunswick	96	2,243,429	$3,704,375
Camden	6	2,308,912	$2,689,144
Pasquotank	8	2,067,876	$2,242,954
Currituck	31	1,952,544	$1,937,133
New Hanover	63	1,693,764	$2,418,878
Perquimans	6	1,623,054	$1,705,59
Chowan	3	1,034,837	$440,952

*Data provided by Brian Cheuront of the NCDMF. The data exclude menhaden and other fish not used for human consumption. Variations in values relative to pounds landed are explained by variations in the types of seafood handled.

Larger operators trade with the larger boats and own a fairly large share of them. Many of these boats fish both in Pamlico Sound and in the ocean. These same dealer/processors also serve the bulk of the out-of-state vessels that offload in North Carolina, if the price is right. In some cases, a continuing relationship exists between the out-of-state fishermen and the in-state dealers. And some of the dealers who own vessels in one part of the state frequently offload and sell their catches to a dealer elsewhere in North Carolina or in another state altogether—that is, again, if the price is right. Medium-sized dealers generally handle products from vessels of under fifty feet that fish inshore, in the sounds, and in protected waters. These companies are oriented toward serving local fishermen whose efforts occur near their homes.

John Bort found the smaller dealers to be the most numerous and heterogeneous. These businesses run the gamut from seasonal sales from backs of trucks to modest, permanent facilities catering to a local retail market or reselling to large dealers. The small-scale dealer may also be a fisherman and will serve other fishermen who use small boats and skiffs. In contrast to large- and medium-sized dealerships, which tend to be involved in national, regional, or statewide marketing channels, the small dealership is more locally oriented.

During my fieldwork in the nineties, I found that that the characteristics of such operations are molded and changed by variations in circumstances, much like the annual rounds of the fishermen. Tourists and summer vacationers, for example, who might have a minor influence on wholesale-oriented large dealerships, can have an important influence on a small operation. Any small-scale operator who has experienced the aftermath of the rash of hurricanes in the nineties—particularly in regard to the drop in tourism, sometimes for an entire season—can testify to this. When we did our research on hurricane impacts in the late nineties, I was amazed at how many small dealers were put out of business. I was equally impressed by how many found the means to survive!

Very few marketers who do not have a permanent facility—peddlers, if you will—derive their total incomes from seafood sales. They hold other jobs, many of which are not even marine related. One of the most successful dealers in this category I have seen is also a public schoolteacher. For these entrepreneurs, seafood marketing is a part-time, albeit important, activity that they need to supplement other incomes.

I became aware of a substantial, general change in dealers' revenue sources during the late nineties. Some of the larger dealers have done very well economically during the prosperous nineties by pursuing a strategy of diversification into other areas, such as marina and resort development, ownership of non-seafood retail markets, and so on. I was impressed, too, with the number of dealer/processors, of all sizes, who were importing and exporting fish. In my view, however, these are just contemporary examples of the continuing line of adaptive strategies.[2] And, as in the past, those who understood the current and near-term trends succeeded. Those who didn't, didn't! Lee Maril, in his study of Texas shrimpers, found a similar pattern had developed nearly two decades earlier in the Gulf of Mexico (Maril, 1983:163).

COMMUNITY LINKS Most fish dealers in North Carolina, more than 80%, were born and raised in the state, and almost that many operate in the counties where they were born. The rest were raised in another coastal state. About one-third of the dealers have fathers who were commercial fishermen, and about one-fifth of dealers' fathers were dealers themselves. Interestingly, John Bort and I discovered that nearly three out of four dealers grew into the business because of previous ownership by family members. When I visited dealerships in the spring of 2003, I found only a few exceptions to this pattern. Most all of the companies I revisited were either headed by the same person I had met with previously or by his son, nephew, little brother, son-in-law, or grandson. It was a little bit spooky and terribly emotional for me, especially in cases where my original respondent had passed on and even more so when the heir to his business closely resembled him.

So dealers, like fishermen, are local entrepreneurs. Even though they are linked to outside markets, the characteristics of coastal North Carolina seafood companies derive from local bases. Culturally, seafood dealers are hardly distinguishable from the fishermen with whom they trade. Many grew up together, sharing the same experiences in the same coastal cultural environment. In fact, my research during the winter of 2002 took me inland to a former dealer/processor/distributor on the Pamlico River who confirmed that this pattern also held in North Carolina's riverside communities. In this regard, the North Carolina dealer/processors we interviewed were, with one exception, different from those Maril interviewed in Texas (1983:163–4). There he found an institutionalized pattern of unethical behavior: namely kickbacks, locally referred to in the industry as "candy." He was not sure how widespread the practice was. I suspect this kind of shenanigan most likely occurs in relationships that are not grounded in a local cultural framework. I think such cultural grounding is an important and distinguishing factor in much of the North Carolina fishery.

The fact that seafood dealers are typically born and raised in the coastal areas in which they conduct business suggests several things about their lifestyles and values. They learn the business from and enter it through friends and relatives, some of whom continue to work for them. Let's take a closer look at this.

Most of the dealers are part of small communities. As lifetime members of their upriver communities and coastal littorals, the dealers have relationships that extend beyond business with other members of their communities. Their lives are stitched into their villages with many threads. They marry local ladies, many of whom were their high school sweethearts. Their aunts and uncles, cousins, and parents all live nearby. Their children attend local schools. They buy their daily groceries from other local merchants. They go to church, weddings, and funerals with fishermen, local teachers, the tax assessor, the owner of the fishing-gear supply house, the guy who cleans his septic tank (I know of no women in this line of work, either—that is, as owner/operators), and so on. This network of diffuse, multi-stranded relationships "complicates" business decisions and actions. For example, regular transactions with a brother-in-law or a revered, aging local fisherman might well be quite different from those one would expect if such transactions were based solely on marketing and supply/demand factors. The reason is that kin and friendship interactions involve a wide range of issues—political, marriage, courtship, housing, child rearing, religious and social activities, and economic interdependence. In contrast, anthropologists and sociologists refer to a *pure* marketing relationship as "instrumental" or "single stranded," existing only for the exchange of goods or services for financial remuneration.

Thus, the issues shaping the conduct of a dealer's operations with local people, including local business partners, are influenced by the relative values placed upon social and economic goals. Not all of his actions, especially those involving fellow locals, are necessarily directed toward maximizing financial gain. The dealer's choice, in many cases, is to forgo that gain in favor of other, more desirable ends. In small coastal communities where business transactions involve kinsmen, neighbors, and long-term acquaintances, variations from patterns that characterize other industries often occur. As we shall see, one of these variations is an interest-free business loan.

SERVICES PROVIDED TO FISHERMEN The role of the seafood dealer is perceived in different ways by the various kinds of people with whom he interacts. To the general public, he is not very visible except in his role as manager of a retail-sales outlet. Even in these instances, he is rarely distinguished from other merchants.

John Bort revealed that the fishermen sometimes present a picture of the dealer as a mix between saint and Satan. At times the dealer is the middleman whom the fisherman perceives as paying the lowest possible prices for his catch. The Commercial Man knows that seafood is expensive in the supermarket. It is easy to perceive the dealer as the middleman who receives an unduly large share of the difference between what the fishermen receive and the store price. Fishermen do not always recognize the real complexity of the marketing chain and the costs of doing business. The dealer, then, is often the vilified scapegoat when market conditions are poor. I found this to be the case when I visited dealers during the shrimp-landings bonanza in the summer of 2002. Even though the fishermen had some inkling that imports were depressing prices, a fair share of the blame was put on dealers.

Bort and I discovered a local folksong that catches the flavor of the tension between dealers and fishermen, albeit only from the fishermen's perspective. It is a bit mean spirited but does reflect how, unfortunately, some fishermen feel and, therefore, is an important part of North Carolina's commercial-fishing history. In the interest of historical accuracy, I've tried to capture the accent of coastal North Carolina in my transcription:

Chapter Seven

THE CAROLINA CANNONBALL
(To the tune of "The Wabash Cannonball")

This train she runs through Kinston, New Bern, and Pamlico,
She runs through Oriental, but she's never runnin' slow.
She pulls into Atlantic, and she makes an awful squall,
Well, hello y'all know this train,
It's the Cahlina Cannonball.
Heah's to Clayton Fulcha, why y'all know this man,
He buys each and every feesh that's swum around this land.
When the run of spots is ovah and flounderin's up this fall,
We'll send 'im back to Baybro on the Cahlina Cannonball.

Now hello, Mr. Fulcha, I'm havin' a time taday;
The people on these Outah Banks, they seem to want moah pay.
Now spots are selling twelve cent and floundah twenty-four,
Ssrimp are selling thirty-five, and still they ask for moah.

[REFRAIN]
Now heah's to the men of Hattras, they work for what they get,
They feesh all day in the boilin' sun, and all they get is wet.
They search the reefs for mullets, and they drag the sound for ssrimp,
When they pay their bills Sadday night, they haven't got a scee-int
   [i.e., cent]!

At other times, the dealer is greatly appreciated, and deservedly so. Besides family, he is the main source for financial assistance, and without his help, making a living would be very difficult, if not impossible. And I can tell you that during my twenty-nine years of research on the state's fisheries, I have met a lot of dealers. As is the case with their fishermen counterparts, they are a tough lot. They work hard, and they are smart. They certainly

Boats of various sizes linked to the Fulcher Company.

are no worse than other any other typical businessmen in a society that measures success by economic power and are, in my opinion, more than a tad better than most. I knew Clayton Fulcher, Jr., personally. In addition to being a fair, tough, and clever man who was dedicated to his profession, community, and family, he was a friend to legitimate inquiries and reasonable management. He never, ever, responded to my requests for time with anything but courtesy and a willingness to help me become more informed about the work he, his employees, and the fishermen and distributors with whom he worked conducted every single day. If there is a more sinister underbelly to his and others' operations, I did not see it.

The one exception was a dealer/processor whose style of competition literally knew no boundaries, and his practices hurt a lot of good people before he was finished. While he was born a North Carolinian, he did not end his career here. He took his game somewhere else, much to the benefit of the Tar Heel State and to the disadvantage of the location where he landed.

Beyond buying their catches, dealers serve fishermen in other ways, including the provision of fuel, dock space, ice, repairs, and fishing supplies.

The range of these services, of course, varies according to the size of the dealership, the nature of the relationship between fisherman and dealer, kinship connections, and so on.

Three-quarters or more of the large dealers with whom Bort and I were familiar in the eighties provided fuel, ice, moorage, and credit assistance. Half or more provided repair facilities and boat and fishing supplies (e.g., gear). Large dealers do so because they serve mostly larger vessels, upon which their businesses greatly depend and which require more services than their smaller fishing-boat counterparts. Only about one-half of the medium- and small-sized dealers provided credit assistance, but most of them still provide ice free of charge. Some of this, however, has changed.

Dealers without a fixed place of business and those with limited facilities do not provide services to the fishermen. Many medium- and large-sized dealers attempt to address, as best they can, the needs of *some* of the part-time fishermen whose harvests they buy. This usually comes with the establishment of a long-term, predictable relationship. For these "partners," fuel and ice are made available at the dock. The ice is free and the fuel is made available at or near cost.

Talking about ice brings to mind a related issue: If there is one complaint that resonates among all of the dealers, regardless of size, it is that the fishermen do not take enough ice with them to assure protection of shrimp against spoilage. Some fishermen may arrive at the dock in the middle of a hot summer to find that their potential dealer "partners" choose not to purchase their catches. One of the conditions of *becoming* a regular partner is to take care of the harvest—to ice it down properly. As I indicated previously, some dealers now express a preference for imported shrimp because of this problem. Christopher Dumas, of the University of North Carolina-Wilmington, indicated to me that he had witnessed a similar preference among dealer/processors in the other parts of the U.S.

However, Susanna Holst, a student at the University of North Carolina, Wilmington, had some rather important, pointed negative comments about shrimp imports. She complained that, in order to create the most suitable habitat for pond-raised shrimp, a great deal of ecological damage is being done to wetlands and mango groves in Asia. Moreover, the Southern Shrimp Alliance is looking into the possibility that some of the countries that ship fish products to us may be linked to *international terrorism*. And we all know how possible support of terrorism became a hot-button issue after the tragedy of September 11, 2001.

Returning to the services dealers provide to fishermen, those dealers located directly on the water keep stocks of frequently needed supplies on hand. Other stores are ordered as needed. These services make good business sense to both parties. They keep the fishermen happy and, most importantly, fishing. If ice is not available, fuel is too expensive, and/or equipment is broken, both parties to the exchange suffer. Some dealers, though not many any more, maintain repair facilities. These range from very modest collections of commonly needed hand tools to well-equipped shops. But like every other business in the post-boom nineties, dealer/processors have tended to scale down activities that do not have a clearly cost-effective outcome.

Dock space ranges from well-arranged slips to mere tie-ups to pilings. Smaller dealers lack the financial capacity to provide anything other than a simple tie-up area. Most do not even have that. And offloading smaller boats does not present the same kinds of problems larger boats do. Thus, the larger operators have made the necessary capital investment to provide and maintain reliable dock space for the easy and efficient offloading of product, sometimes by a machine-driven conveyer, for the larger boats. And those same boats need protection from changing weather and tidal conditions.

Another dealer-provided service is credit. John Bort and I found that dealers devoted an average of about five percent of their working capital to providing interest-free loans to fishermen for the purchase or repair of equipment. Dealers and fishermen were reluctant to discuss this matter in any detail, but we were able to gain some understanding about how the process works.

Fishing is an uncertain undertaking. Fish are unpredictable, as are weather and market conditions, and equipment can and does break down or get damaged during use. Fishermen experience cycles of failure and success. They never know what they have until the net or pot or dredge comes up. Sometimes a trip won't even pay the operating expenses it entails. When times are rough, the dealer's help is crucial.

During my work with him, Bort uncovered three basic credit-service patterns. I was able to substantiate their continuation during my fieldwork in the nineties. But some things have changed, as we shall see. Short-term credit for standard supplies and groceries is routine for continuing "partners." The dealer keeps a record of charges and accounts are settled regularly. Credit is extended when the Commercial Man experiences poor catches over a couple of weeks. Such credit, as I have noted, is standard practice only between fishermen and dealers with a history of doing business together.

"Curly" Smith was a manager of Evans Seafood in "Little" Washington, North Carolina. His company owned three trawlers in the fifty-foot range, and he traveled daily from Washington into communities on the Pamlico Sound to purchase seafood from other producers. His company provided the short-term credit and loans described below to the people upon whom his business depended for a variety of fish and shrimp from the sound. So this practice was not limited to dealers operating only in seaside villages.

In the past, dealers supported fishermen seeking loans from third parties. The prestige and influence of a character reference from a reputable dealer was a valuable asset to the fisherman seeking a loan from a local bank. In some cases I know of, the dealer co-signed notes to back up loans for the purchase of large boats. They appear to have discontinued this practice, however.

Another option is a direct loan from the dealer to the fisherman, and this practice still exists. Such loans are for the purchase of new engines, nets, or major repairs. There were times in the past when such loans would be made for the purchase of a modestly sized vessel. Without this direct-loan option, the fisherman would often be put out of business, and the dealer/processor would lose a source of marketable product. If the fisherman is already in debt to the dealer, it is in the dealer's best interest to get the Commercial Man back out on the water to earn the funds needed for repayment. In some cases, it is to the benefit of both parties for the fisherman to be able to increase his production, and, if the Commercial Man cannot qualify for a loan for a larger vessel and new gear from a lending institution, the only source of assistance is the dealer. Indeed, the larger dealer may *want* to be the lender in order to create the obligatory relationship I will describe below. Curly confirmed to me that this practice was essential to the medium- and large-sized dealers.

In fact, I was able to uncover a situation in which a dealer found himself slipping farther into an undesirable situation from which he could not escape. A crab fisherman needed $5,000 for new pots, which the dealer provided. Soon after the purchase, the fisherman's engine failed. Having already invested in the fisherman, the dealer came through again, this time for $10,000. Then the prices in the crab industry crashed. Almost fatalistically, the dealer moaned, "That's the nature of the business." In other cases I was able to examine, dealers complained that some fishermen do not have the proper work ethic to repay their loans. Repaying loans occurs within a mutually and well-understood framework, even if it is not always verbalized in the legal detail you might expect. Most commonly, the dealer subtracts a portion of the value of the catches, usually 10%, until the loan is repaid in full.

Providing loans acts as another informal means for a dealer to assure himself of continued sources of seafood to buy, sell, process, and distribute. The Commercial Man is under no legal obligation to sell his catch to the lender, but is obliged by custom to do so, nevertheless. The common understanding is that the fisherman will sell his catch only to the lending dealer even if he could get a better price elsewhere. In return, the dealer is obliged to buy the catches of "his" fishermen, even if he could buy the product cheaper from someone else. This mutually obligatory or "prestatory" relationship enhances the security and predictability of both parties in adhering to the informal contract. The fisherman knows he has a source of assistance and a place to sell his product. The dealer/processor knows he has a reliable source of product to take to his market outlets.

As I outlined in the previous chapter, the fisherman has a number of alternative credit sources. But historically (that is, prior to about 1954), other than family, the fisherman had no alternative. Banks considered loans to fishermen to be too risky, and they still do. This often led to situations of extreme dependence on dealers that, today, would be considered exploitative. One fisherman told us, "The dealer provided the fisherman with everything and took everything in return." In some ways, this was similar to land-based sharecropping at the time. The dealer did not control the resource base but did control the capital needed for production, and he was the only outlet for the fisherman's harvest. At that time, more direct sales to the tourist market were, as they often are now, difficult for the fishermen to make.

In the distant past, this strategic position undoubtedly did provide an opportunity for the dealer to manipulate the relationship to his advantage. For example, the low prices many offered to fishermen did not always reflect market prices. But we found that the dealer/processors' perceived advantage had deteriorated as the country entered the second half of the twentieth century. Improvements in transportation and communication—as well as the advent of additional credit sources, such as government insured loans—began to provide fishermen with more options.[3] Fishermen now have a range of available choices if they choose not to accept the services of the dealers. Far more than in the past, fishermen from diverse locations are now acquainted with one another, communicate more frequently, and exchange experiences freely. Thus, dealers now conduct business with better-informed and comparatively independent fishermen. This has leveled the playing field, at least a little bit, and has promoted more equality in the exchange relationship.

OTHER RELATIONSHIPS To get a complete picture of the business framework of the seafood industry, we also need to understand the dealers' relationships with two other groups of fishermen. First, let's examine the dealers' relationship with recreational fishermen, which changed substantially with the Reform Act of 1997.

When recreational fishermen were permitted to use commercial gear and sell seafood, the relationship involved little more than a simple provision of free ice for some, the sale of fuel and food, and then the purchase of whatever product the fisherman would bring to the dock. Those who held such licenses but were, in actuality, serious part-time fishermen would have a more extensive relationship, getting all of the ice they needed, and some would even have a line of credit. I know of one case in which a part-timer received a loan for equipment, but he was a local, born and raised in the remote coastal village, was well respected for his contributions to the community, and was a first-rate fisherman.

"Pure" recreational fishermen were more like hobbyists. And since they had other and regular sources of income, they did not require and would not have received any form of financing from the dealer/processor. Most of them sold to whomever was offering the best price. A lot of them took their catches back to their respective communities and peddled their wares to neighbors, friends, and relatives. One recreational shrimper I knew caught about one hundred pounds while on vacation. He was not satisfied with the offers from a couple of dealers, somewhere in the $2-per-pound range, heads on. The shrimp were "pretty," he told me, about 20–26 count. So he headed the shrimp himself, iced them up, and drove home. After heading the shrimp, he still had more than seventy pounds. He froze about twenty for himself, and gave a number of one-pound packages to family members, leaving him with about fifty pounds. He sold those for $4 per pound, which came to $200. This compares with the $200 he would have received had he sold the entire catch to the dealer and left none for himself or relatives. He more than paid for his fishing trips and groceries for the entire week's vacation. In some cases like this, the recreational fisherman would sell part of his catch to the dealer, keep the remainder and write off the entire trip, sometimes including the vacation as well, as a "business" expense.

With the elimination of the opportunity to sell seafood, pure recreational fishermen have turned exclusively to disposing of their catches on their own, and the exchange relationship with the dealer has evaporated. As I noted previously, the serious part-timers who could qualify for the SCFL licenses bought them. They still may only fish part time, but no longer have to worry about qualifying on the basis of dependence because their initial entry remains as a lifetime benefit until they choose to relinquish it or die.

Large-scale, migratory shrimpers are at the other end of the spectrum in terms of size and commercial activity. Some of these have North Carolina licenses, which permit them to sell their catches to other, distant dealers in the state who are close to where the catches were landed. Out-of-state fishermen want the best prices for their harvests, and when they land in North Carolina, they want markets close to the fishing grounds, as well. They also want the range of services they need to operate efficiently. They require tons of ice; hundreds, maybe thousands, of gallons of fuel; and a market for large volumes of shrimp. They depend heavily upon the consistently reliable services. A shortage of supplies or a delay in obtaining them could be very costly. For these reasons, once the relationship with a large dealer is established, the captain of a migratory fishing vessel returns each year to that dealer when he is working in the area. His ties closely resemble those of the local fishermen, but without the credit assistance normally afforded to the locals.

INTERNAL BUSINESS ORGANIZATION With the exception of one-person operations, dealers need employees with a range of skills. The majority of workers in the processing sector are seasonal, and payments are indexed to the amount of seafood they handle or process.[4] Some workers, not infrequently relatives, have full-time year round positions such as supervisors, bookkeepers, secretaries, etc. With the occasional exception of one long-time employee, food handlers in the collection and processing areas are all temporaries.

The number of employees who work full time for the typical medium- and large-sized operators is four or even fewer. They generally receive hourly wages, but on occasion, a long-term employee receives a salary. The jobs of many employees in these kinds of operations defy classification. They do a variety of tasks, as needed, including offloading, sorting, icing and packing, stacking, and, of course, heading and shucking. Some

Latino dock workers offloading a hefty catch of croakers.

the men have the skills to repair machinery, fishing gear, and take care of other problems that emerge. During one cold winter I observed this kind of help repairing frozen water lines and, later in the year, digging ditches for sewer pipes in a nearby condo development owned by the dealer. In the scallop business, more than once I might add, I've seen operators and their employees construct an entire machine-shucking operation including: adapting a conveyer system to offload product, making a size-sorting box, a steam box to open the scallops, building a sorting box to separate the shells from the animal, constructing another conveyer to transport the scallops to the evisceration rollers, and fabricating a cooling tray.[5]

During the seventies and eighties, the labor force for handling and heading shrimp could increase to dozens in the larger operations. Workers were, and still are, paid according to the number of shrimp headed, as they were when they received a nickel a five-gallon bucket when the industry was first undertaken. Now they're paid per one-gallon bucket for the *heads*, not the bodies, as was the case when the fishery was started.

African-American children heading shrimp.

The workforce used to be drawn from the local community. In the summer, especially, married women and their school-aged children had schedules suitable for work in the processing buildings. It was quite common for the families of fishermen to be involved in processing fish. This pattern provided the dealer with a trustworthy and dependable labor force and supplemental income for the families. This practice occurred not only in the coastal communities, but also in sound-side and upriver locations, as well. And the range of work extended to crab picking (Griffith, 1993) and scallop shucking (Maiolo, Fall 1981). Large dealers in some parts of the state processed multiple species and employed local people for most of the year (Bort and Maiolo, December 1980). But the locals were still called in only when the fish came to the dock, and there were intermittent periods of no work between loads of fish. Some degree of specialization existed, with some dealers concentrating their efforts on a limited number of species. For example, crab, oyster, scallop, and certain kinds of fish processing tended to cluster within the same dealerships. Dealers chose not to handle certain product because of potential scheduling conflicts.

The supply of labor from traditional sources has evaporated in some of the coastal areas. Many Latinos have been recruited to replace both white and African-American workers in some areas, with the exception of Harkers Island in the central coast. This is clearly a consequence of the prosperity of the nineties, which caused acute labor shortages. The workers are almost exclusively Hispanic in some dealerships. In one case, one of the historically largest dealerships in the state, only Hispanics handle shrimp, and they travel back to their native Mexico during the off season. Most of the shrimp, however, even during the season, are simply packed and shipped (heads on) to the new class of entrepreneurs in the industry, the Asian distributors I will discuss later in this chapter.

Other dealerships still employ a combination of white and Afro-American workers who are given steady wage jobs to hold them in place, and they perform all the tasks the dealer feels are necessary. John Bort and

Chapter Seven

Downeast fishermen's wives processing and packing fish in the 70s

I found that the retention rate in the eighties was 88% from year to year. One of the reasons retention has fallen off is that fewer young people have made themselves available for work, especially since the nineties. One dealer complained that he now has to compete with Nintendo games. As they grow up, young people leave their hometowns to obtain a higher education and/or cleaner, better jobs elsewhere.

Another reason is that more young families dissolve and the divorced women with young children, according to one informant, can now receive public assistance that previously was either not available at all or was not considered to be a socially acceptable option. Moreover, the expansion of tourism and the development of retirement communities have created many steady and slightly higher paying jobs (with benefits) that compete for the low-skilled, low-wage labor force (see Maiolo and Tschetter, 1982).

To the extent possible, processing activities that are temporally logical are grouped together: e.g., certain kinds of finfish harvesting and shellfishing occur at the same time of year. Most of the finfish harvested during this period are simply iced, packed and transshipped. Or processors will establish a rotation of activities. Crab picking is most active in the summer, while oyster shucking occurs mainly in the winter months. Some processing activities also tend to be geographically clustered, such as crab picking, shrimping, and certain kinds of finfish harvests in the Pamlico and Albemarle Sound communities.

EX-VESSEL SALES AND MULTIPLE MARKETS When the fisherman reaches the dock, the exchanges are straightforward. The catch is weighed and payment made according to the prevailing rates, which can change frequently. Once shrimp are headed and packed, the next step in the marketing chain occurs in a variety of ways. The product can be sold directly to retail customers and restaurants or passed on to other wholesalers, processors, or distributors. In the past, some dealers intentionally excluded themselves from certain markets while emphasizing others. For example, while one dealer would not sell even a minimum amount of his products to retail market outlets because he preferred to deal only with other distributors or processors, another would do just the opposite. This is another feature that has changed.

Alternate markets are insurance policies. The degree to which a dealer can cultivate them directly affects his business to the same degree as his success in insuring the supply of product from "his" fishermen. When we first began studying the shrimp industry, it was clear that the larger the dealership, the more likely the owner would operate in a number of markets. Bort and I found that virtually all of the large operators wholesaled shrimp to others, while less than 10% moved to retail outlets. Between 20 and 30% engaged in other sales patterns, including restaurants, brokers, and processors. Medium- and small-sized operators did more overall and shrimp-specific selling to local outlets and retail companies.

For most of the last part of the twentieth century, just under half of North Carolina's shrimp found their way each season into out-of-state markets. Bort was struck by variations in sales patterns as they linked to these different kinds of outlets. He felt that there was a two-edged marketing strategy in the statewide seafood industry. The large dealers handled large volumes of products. They needed, and still do, secure markets for large quantities of perishable fish. These markets still do not exist locally, so large dealers' attention focuses on large-scale buyers elsewhere. Retail sales used to be almost incidental to the large dealer's overall operation and were mostly to restaurants because of the large volumes they need during the tourist season, not to mention the consistent product quality and the certainty of supply that they require. This has changed somewhat because of the growth in the tourist industry.

The medium- and small-scale dealers have always been in a better position to engage in direct retail sales. Local consumers and tourists began to present a more lucrative marketing opportunity, especially beginning in the 1950s. The smaller volumes of seafood such dealers handle were more easily and readily absorbed by the retail markets. And while the size of each transaction was smaller, the profit margin was greater per unit of product. This continued until the mid-eighties, when imports became available.

One of the interesting things we uncovered in our original fieldwork was a relatively common practice of assuring prime customers a wide range of products by purchasing from other dealers, if necessary. So, even though a given dealer might prefer to handle just a few products, such as shrimp and one or two others, he will expand his activities, if necessary, to secure the relationship with those to whom he wants to sell. Following this practice seemed to be more likely the larger the dealership. I suppose this is why such dealers grew so large and became so successful in the first place. Medium-sized operations would engage in this practice, as well. Let me expand on this.

Initially, this pattern of buying and selling was undertaken primarily as a service to customers rather than as a reflection of personal preference or as an opportunity for potential profits. A dealer's primary product may be shrimp, but if he chooses to sell regularly to restaurants, he will also often provide his customers with scallops, oysters, clams, and so on, even though his sales volume may be limited and his profits marginal. In order to do this, the dealer has to buy some of the other types of seafood from other dealers, wholesalers, and/or processors if he were not able to buy and handle the required species from the fishermen at the dock, or if he simply didn't want to bother to do so. This seemed to explain why we found some medium- and large-sized dealers handling small quantities of a number of certain varieties of fish in addition to their primary product(s).

Curly Smith, former manager of Evans Seafood (whom I introduced earlier), emphasized the importance of cultivating and maintaining good relationships from one end of the marketing chain to the other. For example, it is important to have the cash available to pay the shrimper on the spot at the dock. But, on the other hand, Curly had to assure his other suppliers and buyers that their credit was good. And this would reveal itself in a variety of ways. In one case, a struggling dealer, let's call him Fred, from the Beaufort, North Carolina, area was a valuable connection into other supply networks, particularly in South Carolina. So Curly would take thousands of dollars in cash, and he and Fred would head into the distant coastal docking areas in the Palmetto State. Curly would put up the money for Fred to buy the fish they couldn't get in North Carolina, then buy it from Fred on the spot, giving him a profit, and bring the purchased product back to "Little Washington" for distribution to Curly's customers. And he would often do this just to keep the people who would buy *from* him satisfied so they would continue purchasing the shrimp and other seafood that he could get in large quantities locally and wanted to move out through them. Thus, this type of transaction serves two goals.

Because of the price variation of shrimp, which is tied to their size (count), they tend to be channeled into different markets. Even before the influx of foreign shrimp, there was a general belief among dealers that the controlling market factor lay outside of the state. The price was affected by what the "breaders" in Georgia and Florida were willing to pay. These breaders—for example, the King and Prince Company in Brunswick, Georgia—are large operations that process and freeze shrimp for national distribution. Today the large distributors who import and sell processed shrimp to large retail and discount outlets have an equal or larger influence on prices, as I revealed in the previous chapter.

Historically, breaders wanted the largest shrimp and only a few other potential buyers attempted to compete with them because they offered the best prices. The notable exception occurs when dealers make direct retail sales of larger shrimp to consumers because this offers a better profit margin than selling to breaders at wholesale. For local retail outlets and restaurants, buying the same product through another channel would have put them at a competitive disadvantage with others that could serve up smaller shrimp for a smaller tab. Most lower-priced restaurants still prefer to buy the smaller shrimp because they are cheaper. Also, they can serve more shrimp (in terms of numbers) on each plate and, by breading them, make the meal look even larger. Breading constitutes one-third to one-half of the product on the plate. The pricier, upscale restaurants want the larger animals to steam, broil, or sauté for their gustatory presentations, and, of course, they charge accordingly. But the people who dine in such establishments know in advance that they will pay more for their meals. The patrons of traditional coastal restaurants serving traditional southern fried seafood tend to be more price conscious.

The next time you are in the Morehead area, stop by the Blue Jeep shrimp dealer on Route 70, just east of Route 24. The business literally started out of the back of a light-blue Jeep parked along side of the road. The operator, I'm sure, has owned several of them since he first opened, but he keeps one sitting beside the building he now uses for his business because it has become a sort of logo. For years, people have traded with the company, either on their way through the area on a vacation or on their way home. In either direction, it is an easy pullover, and you can get an accurate idea of the differences in prices as they relate to size. You will be able to purchase all sizes, from large shrimp down to small bait shrimp, with the heads on or off.

Larger dealers don't like to talk about their contacts with markets outside of the region. Bort referred to such contacts as trade secrets. A lot of time goes into cultivating and securing relationships with buyers from other areas, and the dealers are not about to compromise them. According to Bort, revealing a list of customers is like inviting competitors to steal your best business channels. The dealers, however, are willing to discuss the general aspects of their marketing channels, and some things sure did change during the last quarter of a century.

But before looking at those changes, I need to mention one thing that did become very clear to us and which hasn't changed: the amount of trust that must prevail between and among these business partners, *some of whom have never met face to face!* In conversations with several dealers with whom I have had extensive discussions, I was dumbfounded by the amount of business that occurred, and continues to occur, over the phone. Thousands of dollars can be involved in an exchange between two people who may have no idea what the other even looks like. That kind of trust is secured by years of doing business with each other and exchanging hundreds of thousands of dollars in transactions without a hitch. Some of these take place between people separated by hundreds or even thousands of miles.

Relationships with distributors, processors, and middlemen are critical to the continuing success of the North Carolina seafood dealer. The volumes and types of seafood available to the operator vary. Markets vary as well, as supplies from all sources range from scarce to glut, as landings and import data show. Demand varies, as well, based on availability, price, and season. The dealer must be able to access a range of markets, from local to distant ones. Otherwise, he might find himself holding a large quantity of rotting fish or facing an order for two thousand pounds of seafood that he does not have. What is interesting is the strategy dealers have chosen in order to deal with this problem, particularly in light of recent events. A number of them began to buy imported shrimp and crabmeat from national distributors in the late eighties and early nineties and then move the products through their marketing channels.

So, as we have seen, opportunity, location, and family history all play a role in someone getting into the dealer/processor/distributor business. Historically, the operator has had at his disposal an abundant supply of seafood that would be of little value if it could not reach the consumer. In almost every case in North Carolina's commercial seafood industry, the supply of the various types is far greater than the amount that can be absorbed into local markets where the products are landed and dealerships are located. This is still true, despite the incredible expansion of permanent and recreational populations in and around fishing communities during the nineties. Therefore, non-local markets are essential to the industry. By the same token, the North Carolina dealer has learned that his business' revenue cycles can be made to run more smoothly by opening channels for purchasing an even wider range of seafood from other distributors rather than just from fishermen. All of this occurs within an exchange system that rests heavily on trusting relationships that are built over time.

Some of North Carolina's seafood products continue to move directly to local markets. Others still move to the state's inland markets but no longer do so through the traditional chains. But, as was the case with the dealer's labor force, the nature of the ethnic mix in the marketing chain has changed dramatically since the mid-eighties. The amount of product moving to breaders located further south has decreased as a result of competition from imports. This decrease has been compensated for, to a large degree, by some large, and even medium-sized dealers' sales—as much as 80% of their shrimp—to what they label "Asian" middlemen. These middlemen, in turn, sell to other Asians, some of whom are Asian Americans, who then sell to Asian-owned retail outlets and restaurants. What is even more interesting is that a number of traditional dealers have begun to buy imported shrimp and crabmeat from a national distributor and then move the products through this comparatively new marketing chain.[6] The major out-of-state recipients of North Carolina shrimp are along the East Coast, from New England to Florida. Historically, Norfolk, Virginia, New York City, and Tampa have been the most important out-of-state secondary markets. A Canadian market for finfish, especially reef fish caught in the southern part of the state, developed in the 1980s, according to one of the informants I interviewed in 2003.

New Orleans was a destination for some of the seafood distributed by the Evans Company in Little Washington. From New Orleans, the product was dispersed to retail outlets, either directly or through other wholesaler/distributors. And this brings us back to the topic that we discussed earlier in the chapter—the complaint that the dealer can easily manipulate prices.

Multiple markets affect prices in a number of ways. With a lot of the shrimp finding its way to out-of-state outlets for further distribution, prices inside the state are affected by competition from distant markets. Some of those markets have access to other domestic product and imports, as well. And, as we have seen, imports from foreign producers have had an incredible impact on prices in real terms. The marketer, then, is in the middle of a framework over which he has no control. If he cannot sell his goods in distant markets at competitive prices, he loses those customers, even if only temporarily. When that happens, he has to increase local sales of the product, which, in turn, depresses prices there. Those markets, too, are now in the crosshairs of importers, mainly through the large discount retailers that have found such locations to be quite lucrative. Either way, the marketer can only offer a price to his fishermen that will allow him to sell to someone else at a profit. Otherwise, he has no reason for being in business. A seafood dealership is a tough business, and it takes a tough person to run one.

To transport shrimp and other seafood products to distant markets, North Carolina's seafood dealers initially attempted to use the rail system, as I mentioned previously (Maiolo and Tschetter, 1982). It didn't work. Today some North Carolina companies specialize in shipping seafood by truck. Independent haulers often move shrimp between in-state and out-of-state wholesalers. In a lot of cases, dealer-owned refrigerated trucks typically transport boxes of shrimp. In other instances, buyers send their own trucks to the dealers to pick up the product.

Through the years, several of my respondents indicated that some North Carolina truckers who delivered seafood products to the Baltimore, Philadelphia, and New York markets faced difficult situations. The men who offloaded the products, according to reports, apparently expected that a portion, five pounds per box, would be set aside for their personal use or sale. I have not been able to confirm this, but a couple of trusted respondents told me that when some truckers did not cooperate, their vehicles were burned, and they were not allowed to return. The trucker faced a dilemma: if he gave the guys on the loading dock what they wanted, the buyer would complain that he had been "short weighted," even though he knew what the source of the problem was; if he didn't, he might have to deal with a very real threat of violence. One solution was simply to deduct the difference from his invoice to the buyer.

An interesting sidebar here is the apparent change that has occurred in Asian buyers' demand for a certain kind of shrimp. You may recall from the first chapter that, historically, Asians, especially Chinese restaurateurs, have preferred white shrimp (greentails). Operators tell me that quality and price now prevail in the Asian market segment and that the species does not matter. The reason for this change seems to be rooted in the growing popularity of ethnic foods, in general, and Asian cuisine, in particular, among Americans. Because of this, the demand for white shrimp may have exceeded the supply. Recall, too, the seasonal nature of the white shrimp in contrast to the availability of other species and imports throughout the calendar year. Moreover, the competitive nature of the Asian restaurants, along with the growing number of lower-priced and fast-food restaurants in this class surely have combined to enforce a buying pattern more consistent with competitive prices.

Now let us move on to the last chapter for an examination of the cultural and community contexts within which the production and initial distribution systems of the shrimp industry operate.

The Struggle for Survival in a Changing Coastal Culture

The divisions and animosities that continue to prevail among the stakeholders in the seafood industry, in general, and the shrimp fishery, in particular, are many and unfortunate. Since first beginning my fisheries research in the seventies, I have marveled at the lack of understanding among the parties who may have different specific objectives, but who surely all have a stake in the health of the state's marine resources.

Each party to the industry may have legitimate complaints against the other. But sometimes the seeming unwillingness to understand the needs of the others is maddening. This is coupled with, and complicated by, the fact there are those who do, indeed, put personal motivations above what they know is the right thing to do for the fishing industry. And I'm not just talking about fishermen and dealers. For example, while I was working with the various management groups, I had the opportunity to attend many receptions, informal meetings, and public sessions. At one of the receptions, I was having a discussion with a member of a management body when the chairperson of a subcommittee interrupted us: "Excuse me" he said, "I need to talk to Pete [not his real name, of course] for just a minute." Right in front of me, he then said, "Pete, tomorrow at the meeting I am going to vote against the proposals to regulate the [blank] fishery, but you know I support them, and I want you to vote for them.

Otherwise, the damned commercials will have my ass, and the Governor will take away my seat [on the regulatory body]."

Scientists are not without fault either. On occasion I have seen their egos stand between them and the correct scientific analysis of an issue. I have also seen them fail to take the needed time to recognize the enormous impacts their recommendations have on fishermen who can hardly get by on their scant incomes. Some commercial fishermen cannot see beyond the bows of their own boats, and they, along with a few dealers I have known, support regulations so long as they limit someone else. As for the recreational fishermen, I have seen successful businessmen attend meetings and complain about the "large" amounts of king mackerel allocated for commercial fishermen to harvest, while they, the successful businessmen/recreational fishermen, were filling up their coolers and selling off their catches for revenue they did not need.

But all of this constitutes the reality, and it probably isn't going to change. We live in a society where the pursuit of gain at the expense of others is expected, sanctioned, and even admired. The good news is that ours is a pluralistic society, one in which a number of interest groups can pursue and then protect their interests within an organized, albeit cumbersome, political framework. This is the kind of framework within which the fishery management's access and allocation decisions are made.

It may not be perfect, but the Magnuson Fishery Conservation and Management Act, first implemented in 1976, is an effective nationwide management system. It assures that every special interest will be represented at the decision-making table, even if it means a given interest group may have more representation than another. The provisions of the Act specifically charge regional councils to take into account historical fishing practices and enjoin them not to govern solely on the basis of biological or economic issues. This isn't always the way things turn out in practice, but it sure is better than the alternatives.

Chapter Eight

Knowing what I know about the dynamics of the fishing industry, the culture within which it operates, and the state, regional, and federal management frameworks that surround them, makes me marvel at what I call the "Miracle of '97." If the state management system developed in the sixties for the shrimp fishery was one of the first models for the best use of science, surely the passage of the Fisheries Reform Act is the model for the best use of the political process. The intent of the Act is clear to me: to limit, if not eliminate, the participation of those with outside, non-fishing sources of income from unfairly competing with those whose livelihoods depend almost entirely on fishing. The trend toward the "gentrification" of fisheries is driven by increasing pressure from part-time fishermen using commercial gear, which the regulators felt would eventually lead to proposals to professionalize, even limit, entry into the commercial fishing industry (Orbach and Johnson, 1988). They were dead-on with their prediction. Too many retirees and weekend or seasonal fishermen had been creating havoc in the harvesting sector, conflicts among user groups, and damage to the fish stocks. This led to the climate in which the Reform Act provisions became attractive to all parties involved in the problem. I would like to note, also, that the team of Johnson and Orbach (1996)—along with David Griffith (1996) and Barbara Garrity-Blake (1996)—played a key role in the development of the social-scientific justification for the creation of the 1997 Fisheries Reform Act.

I want to delve deeper into the cultural framework that sits beneath the various parties involved in the shrimp fishery, in particular, and describe what I believe are important changes that have taken place. What we have historically known as the "coast" is changing, both physically and culturally, in North Carolina, as it is elsewhere. Much of that with which we are familiar is disappearing, and we can expect more of the same in the future. I am not going to take the view that this is bad—or good, for that matter. The lives of many people who live in coastal communities have improved tremendously, but not without a cost. Among the results of these changes is that a way of life is slowly dying out and needs to be chronicled before it disappears forever.

THE GEOGRAPHY AND DEMOGRAPHICS North Carolina's coastline can be divided into three separate geographical areas. Let's examine them one at a time.

To The North The northern section of the coast is best known for its Outer Banks, a series of elongated, sand islands extending from the Virginia line at Corolla down to Ocracoke. The Core and Shackelford Banks are often included under the "Outer Banks" rubric and are regarded as the two southernmost of this chain of islands. But I am going to include them in my discussion of the Central Region because I feel that they are more culturally and sociologically tied to that section of the coast.

These narrow strips of sand, also referred to as barrier islands,[1] are formed by the action of currents on coastal geological features and fence out the direct impact of the Atlantic Ocean. In 1887, Earll (p. 447) took note of the "outer bars," describing them as "bald ridges of drifting sand, almost destitute of vegetation. Owing to this fact they have few inhabitants." I wonder what Mr. Earll would make of things there now. Like their counterparts to the south, the islands and the nearby mainland constitute a multiple-use zone that includes both permanent and temporary (vacation) housing, a variety of recreational uses (boating, fishing, swimming, etc.), and a variety of commercial uses, of which commercial fishing is the most recognizable (Maiolo, 1994:1). The islands are where the beaches are, and they used to be described as a "long way from anywhere."

Sir Walter Raleigh established the first colony (later known as the Lost Colony) on Roanoke Island, which stands between the Outer Banks and the mainland.[2] The Lost Colony Theatre is one of the major visitor attractions in the region. Another is the site of the Wright Brothers

Chapter Eight

National Memorial, the location of their first flight on December 17, 1903. The nearby waters became known as the Graveyard of the Atlantic and have been memorialized in David Stick's book of the same name (1958). He chronicles a long list of ships that went to the bottom while trying to navigate the tricky Diamond Shoals, along with the ships sunk by German submarines in both world wars. This is, as well, the resting place of the famed sunken Union ironclad, the USS Monitor.

Travel into the area has always been difficult. Visitors from all directions must navigate through what used to be many miles of rural villages that are becoming larger communities with more and more shopping centers and traffic. Ferry travel, developed decades ago, still carries passengers from Cedar Island or Swan Quarter to Ocracoke and then from Ocracoke to Hatteras. The flow of people and materials between the mainland and the Outer Banks has always been restricted by the region's low elevation. The construction of roads and bridges to facilitate traffic has been slow in coming, and even after they are completed, they are easily damaged by hurricanes and other storms. Because of this, communities in the region were, in the past, largely insulated from change. Their residents were either fishermen or employees at the lifesaving stations (precursors to the U.S. Coast Guard, established in 1915).

Owing to the historical isolation of the region, its emphasis on recreation, and the general absence of farming (other than for subsistence), the resident African American population has been, and remains small, currently consisting of fewer than one hundred. The percentage of African Americans increases inland. Dare County has provided commuter jobs for some of the inland African Americans in the fish-processing industry, especially in the scallop fishery.

The impetus for an improved transportation system came when the region, particularly Nag's Head, came to be seen as a desirable vacation destination. This first occurred in the mid-nineteenth century when wealthy landowners used boats to get to the islands. The growing popularity of the area eventually prompted the construction of a rail line. In the 1920s, however, powerful interests led to the construction of roads and bridges, and the region began to flourish economically. Interestingly, one of the first bridges to be built during this period, linking Roanoke Island to Nag's Head, was constructed to satisfy the needs of Wanchese fishermen who wanted to drive to the beach to set nets in the surf (Maiolo, et al., March 1995:8). The very important connection to the west, between Roanoke Island and the mainland, did not come until 1955. Another important link was made when the Herbert C. Bonner Bridge was built over Oregon Inlet to help link up Hatteras Island with the mainland. These bridges, along with an improved ferry system, provided accessibility for large numbers of people and increased commerce to and from the mainland and between the islands. As a result, the economy in the region has changed from one primarily dependent upon natural resources to one that is increasingly dependent on the service sector, with one notable exception.

The village of Wanchese, well off the beaten track on Roanoke Island, still remains dependent, to a great extent, on commercial fishing. Harbor development and channels dredged in the 1950s facilitated the use of larger commercial fishing vessels. This was followed by the local construction of steel-hulled vessels, as well as large sportfishing boats. The local fishing industry experienced a boom at the time and peaked in the seventies. When Oregon Inlet became less navigable in the eighties, commercial fishing began to decline, but still remained extremely important (Maiolo, Petterson, et al., September 1993:61–62). Many of the Wanchese boats and fishermen began to fish in distant places—from New England to the coast of Florida and currently as far away as Venezuela.

Fishing interests benefited from improved road transportation, as well. With the new roads and bridges, along with the improved harbor in Wanchese, processing houses in the area could now truck their products to distant markets. And charter businesses, along with marinas servicing private fishing boats, began to flourish. Surf fishing in the area has become one of the most popular activities for tourists during the spring and fall of each year. By the mid-nineties, the number of visitors to the Cape Hatteras National Seashore alone was over two million (Maiolo, et al., March 1995).

The Outer Banks are now a destination for vacationers of all types—daily, weekly, seasonal—and the area has responded with the appropriate construction of motels, restaurants, and cottages. It has been estimated that, in 1990, one-third of the housing units in Nag's Head alone were designated for vacation use, and that number radically increased during the economic boom that followed (Maiolo et al., March 1995:13 passim). By the year 2000, the number of short-term vacation rentals had grown to over 13,000 in Dare County alone. From data I have been able to sort out, the number of vacation rentals on Ocracoke Island is somewhere in the 600-unit range. Research we conducted in 1995 indicated that 75% of the residents of forty-one counties in eastern North Carolina visited the coast at least once a year and that 25% visited between three and eleven times annually! The Outer Banks region was second only to the Morehead City area as a visitor destination (Maiolo, September 1995).

One of the intriguing cultural developments in the area—including the Dare County seat, Manteo (on Roanoke Island), as well as the other islands—is the rotation of tourist-oriented activities, a kind of annual round, if you will.[3] I am referring to the changing rhythm of activities, much like the fishing activities of the Commercial Man. Many of the villages in Dare County (from Hatteras to the north and west to Manteo) and Ocracoke in Hyde County (south of Hatteras) change from sleepy, fishing-oriented communities in the late fall and winter to bustling tourist destinations from the late spring and into the early fall; they then reorient themselves toward fishing during the next late-fall and winter months.[4] It is estimated that the average daily population of Dare County alone swells to more than 200,000 in the period extending from June through August.[5] Labor-force characteristics change accordingly, with high school and college students converging on the area to work as service personnel. Low-skilled Afro-Americans and, recently, some Hispanics work in the area as commuters. It is too expensive for them to live there.

Dare County has become a popular destination for retirees and other people looking for a pleasant, coastal lifestyle. From a mere 7,000 permanent residents in 1970, the permanent population grew by more than 300% by 1990 and another 32% during the nineties, to nearly 30,000, increasing more than four times in thirty years. By comparison, North Carolina, as a whole, more than doubled its population from 1970 to 1990 and increased its population another 21% during the nineties. The nation, as a whole, grew by 13% during the nineties (Maiolo, et al., March 1999; Whitehead, et al., 2000). The state's number of residents over sixty-five years of age grew by about 2% in the nineties, slightly over the national average. During this same period in this region there was an influx of very successful *early* retirees who constructed homes the likes, size, cost, and quantity of which had not been seen before.

There is a dichotomy between long-time residents (let's call them locals) and the new arrivals. Visitors and the new residents are appreciated for the revenues they bring with them. But relationships between locals and seasonal visitors are mostly single stranded, i.e., related to the economic-exchange relationship. There are some exceptions to this, such as the seasonal visitors who own property on the islands and come back each year.

The mutual isolation in the winter months tends to break down some of the social distance between the locals and newly arrived permanent residents. The likelihood of this occurring, however, depends a great deal on the appreciation shown by the new arrivals for the history of the area and their willingness not to project themselves as agents of change. If there is one thing that uniformly rankles locals along the entire coast, it is referring to "how we do things where I come from!" Another is referring to how "slowly" things get done, this by the same people who tell

Chapter Eight

friends and neighbors "back home" how charming the lifestyle and pace are in the South.

In 1995, I discovered something most interesting about the values and orientations of key opinion leaders regarding future development on the islands (Maiolo, September 1995). I had expected to see resistance, especially among some of the long-time network of influential individuals. What I found, instead, was that, while there is a great of value attached to both the natural and human environments, there is also concern that the area should continue to develop. Furthermore, the leaders did not see such things as improvements and expansion of the transportation infrastructure system, which would bring more people into the region, as conflicting with their cherished environmental values. This is a very important factor in the recent development of the *general* area. I had also become aware of another matter when it came to coastal development. Ellis (1986) informed us that small communities may look alike on the surface but might not be as similar as first impressions would lead one to believe. She prompted me to look more closely at how communities differ in their historical circumstances, the dominance of certain community institutions, and their connections with other communities. These, in turn, reflect differences in how residents see the future of their villages.

I did, indeed, find subtle differences among the Outer Banks villages. Emphasis on development was lower, and status-quo thinking more widespread, in the Ocracoke and Hatteras/Buxton areas, where development has not been as rapid. Communities in and around the Nag's Head area supported and experienced higher rates of development in the nineties (Maiolo, September 1995:39). I had already found that specific revenue and employment sources vary by community. For example, as I noted, on Roanoke Island, Manteo serves as the seat of government, and does depend heavily on tourism, while Wanchese is predominantly a fishing village. Most of the businesses there are tied into or are dependent upon fishing (large commercial trawlers with crews, boat builders, seafood packing, etc.). As one moves southward down the islands from Nag's

Head—where the principal source of revenue is recreation, while leisure and recreational activities, including sportfishing, remain very salient—the presence of traditional commercial fishing becomes more visible. This is especially evident at certain times of the year, particularly in the lower Hatteras Island communities and Ocracoke.

Our work in the early nineties revealed that "Despite the fact [that] ... the commercial fishery in Ocracoke is relatively small, ... [residents] ... in this community consider this to be one of the common uses of the environment" (Maiolo, Petterson, et al., September 1993:20). The community, however, is perhaps best known for its historic sites, extensive beaches, recreational fishing, and antebellum charm, which attract thousands of visitors each year. It still is accessible only by water and air. The remaining pristine coastline here and to the north has been secured by the National Park System's administration of the resource. The same is true for Core and Shackelford Banks, both of which are protected from any form of housing or business development that is not already grandfathered in.

So, while the commercial fishing culture remains as part of the economic mosaic of the region, its influence is clearly on the wane. We will see that the northern coastline is not the only place I have found this to be true. Such a trend, as you can imagine, makes for a declining political constituency representing traditional fishing interests. As the development ethic gets more traction, and it will, less and less of the region will be viewed through the prism of the Commercial Man. How and how quickly a place like Wanchese will capitulate to this trend, if at all, will be very important. I have always felt that if you want to know whether some of the important historical commercial traditions are or are not in danger of extinction, look to see whether efforts are being made to preserve boat building, gear-assembly companies, older marinas and fish houses as they become surrounded by beach houses on stilts, motels, condos, restaurants, bookstores, gift shops, and stores that sell souvenir T-shirts. Usually, that will tell the story.

Chapter Eight

The Central Coast With some differences, this part of the coast is a repeat performance of the trend in the region to the north. For example, while the population of Dare County was growing by 300% during the final thirty years of the twentieth century, centrally located Carteret County grew by 89%, from under 32,000 to about 60,000. But, in many areas, that growth induced just as much economic and cultural change. By the year 2000, the number of short-term vacation rentals had grown to more than 13,000, nearly identical to their number in Dare County.

Morehead City is the center of commerce in the central coast and is the fastest growing municipality in Carteret County. About seventeen miles to the west sits the Cherry Point Marine Corps Air Station, established in 1941, with a detachment of 20,000 personnel and a civilian workforce for the Naval Air Rework Facility also located there. The base is intricately tied into the Carteret County economy, providing hundreds of jobs and producing thousands of tourists. And it is not uncommon for those stationed or working at the base to live throughout the county, including the down-east fishing villages.

Morehead's influence also extends as far south as Swansboro, which actually is part of Onslow County. In nearby Cape Carteret, the Marine Corps also has Bogue Air Field to support training exercises for the Cherry Point and Camp Lejeune (Jacksonville, North Carolina) bases. Morehead City is home to one of North Carolina's two deepwater industrial ports, made possible by Beaufort Inlet, which provides passage to and from the Atlantic Ocean. Across the bridge to the south sits Bogue Banks, an island that runs east to west and home to thirteen miles of beach resorts and retirement communities. The island is the site of another very important historic landmark, Fort Macon, which was built during the early-nineteenth century. It draws thousands of visitors each year.

Many of the early white settlers to the Morehead City area either transferred their living arrangements from Core and Shackelford Banks, emigrating after the devastating hurricanes of the late nineteenth century;

or came from the inland to the west, following the development of railroad lines to the coast. Later, African-American settlements provided cheap labor that aided the area's growth, including the development of the commercial fishing industry. Today, Afro-American residents represent 20% of the permanent population. Most of the jobs they hold in the fishing industry still are of the low-skill variety.

With its incorporation in 1861 by Governor John Motley Morehead, the city was destined as a hub for commerce and economic diversity, of which the port was to be the centerpiece. In addition to its current import/export role, the port now serves as an entry and exit point for U.S. Marine Corps troops and equipment from Camp LeJeune. Following the Civil War, the city became a central distribution point for commercial fishing products, and, after World War II, it became the sportfishing capital of the central coast (Maiolo and Tschetter, 1982). Both of those activities continue to be important today. But Morehead City has also had a long history of family recreation, particularly in conjunction with its neighbors, Atlantic Beach and the other Bogue Banks communities just across the bridge. The island neighbor is now home to many of the more than 13,000 vacation cottages and villas I mentioned previously. In addition to the hundreds of hotel rooms and thousands of cottages and condominiums, there are marinas with huge dry-stack facilities for recreational boats, dozens of T-shirt and souvenir shops, shopping centers, and dozens of every type of restaurant and nightclub. A typical summer day swells the island resort communities' populations from under seven thousand to several hundred thousand. Literally in the middle of all of this is Salter Path, a village populated by a few hundred descendants of squatters who moved there in the early-twentieth century. Their formerly quiet, out-of-the-way fishing village is now surrounded by the sights and sounds (and congestion) of tourism. Much the commercial fishing, dealer/processor, and restaurant activity is now directed towards the islands' recreational presence. I might add that one of the best seafood sandwiches you will ever experience can be found at the Big Oak drive-in there (or at Els, on Route 70 in Morehead).

To the east stands Beaufort, the county seat, which was finally connected to Morehead City by a bridge in 1928. It has the second largest population in the county, about four thousand, and is a major attraction for tourists. The view from the waterfront is nothing short of spectacular and includes the Rachel Carson Preserve, which is home to the famous wild ponies. Originally inhabited by Native American tribes that were later displaced by English and French settlers, the Town of Beaufort was established in 1713 and incorporated in 1722. During the eighteenth century, it became home to whaling activity on nearby Shackleford Banks, shipbuilding, and international trade. It was known then as Fish Town. From there, meat, furs, fish, timber, turpentine, and pine pitch were shipped (Maiolo and Tschetter, 1982). As a port, Beaufort was also of strategic military importance for privateers at several points in its history and for the British during the American Revolution.

As early as the mid-nineteenth century, Beaufort had become a resort destination, as well as a center of maritime commerce. But, as the deeper and more accessible Morehead City port developed, Beaufort lost its status as a location for merchant trade. It did become an important distribution center for fish products, especially menhaden, which dominated the local economy for much of the twentieth century (Garrity-Blake, 1994). The eventual construction of a bridge to Morehead City and a rail line greatly facilitated overland trade. Across from and in full view of the waterfront, the community eventually, and unfortunately, became a town of rotting wharves and fish houses. Urban-renewal funding revivified the waterfront in the 1970s and transformed much of the community back into a destination almost exclusively for tourists. At the same time, there has been a great deal of emphasis in the community on refurbishing and preserving the historic buildings through the town's historical association and its town board.

From the standpoint of commercial fishing, Beaufort is still one of the two remaining communities on the Atlantic Coast where menhaden landings and processing still occur. That industry has always been a source of employment for African Americans, who comprise about 22% of the community, and for the nearby unincorporated villages. Unlike other fisheries, which have been or are undergoing changes in their labor forces, the only change in the menhaden industry has been the residential location of the boat and factory workers. In the past, they came from homes in the down-east village of North River. At present, the workers come from Harlowe, a community to the west of Beaufort.

Boat-building and repair facilities still exist there, as do a few seafood dealers. And a dozen or so trawlers and snapper/grouper boats still work out of that community. But the real ballgame is tourism, with the historic waterfront, museum, upscale restaurants, and bed-and-breakfasts. As boats approach the final channel leading to the community, the Duke University Marine and National Marine Fisheries Labs appear on Piver Island as sentinels guarding the marine resources in the area.

In Morehead City, Beaufort, and along the Bogue Banks communities, the trend that Orbach and Johnson wrote about in the eighties has continued: "What is clear is that the fishing industry is only a small part of an increasing[ly] expanding and increasingly complex coastal environment" (Orbach and Johnson, 1988:66). What is more impressive is how this kind of change is occurring in remote places where, even fifteen years ago, it could not have been imagined. An examination of the "down-east" communities of Carteret County will justify this assertion.

The Down-east Connection In 1982, Paul Tschetter and I wrote that while tourism was flourishing in diversified Carteret County and having a unifying effect on it, the "growing interdependence ... has not been uniform" (Maiolo and Tschetter, 1982:209). Even though the early settlements spread in a northeasterly direction through Beaufort and on "down-east,"

Chapter Eight

it was the development patterns to the west of Beaufort that gave Carteret County its economic pluralism during much of its history, as we have seen. This began to change after World War II, and took quite a leap during the prosperous nineties. I will chronicle how this happened through a discussion of two of the many charming and historic villages—such as Bettie, Otway, Marshallberg, Williston, Stacy, Sea Level, Cedar Island, and others—on and near Route 70 from Beaufort to Cedar Island. The two I have chosen are Harkers Island and Atlantic.

Originally known as Craney Island, Harkers Island was first inhabited by the Coree Indians, for whom Core Sound was named. Mr. Ebeneezer Harker acquired the one-by-four-mile island in 1730. When the 1899 storm hit Diamond City on Shackelford Banks, some settlers moved to Harkers Island. Claiborne Young has noted, "These skilled whalers brought an intimate knowledge of boats and seamanship with them, and their 'sea-sense' lives on to this day [in the island's boat builders]" (Young, 1993).

Boat building and commercial fishing became the most important economic activities on the island and remained so through the mid-twentieth century. Then two motels were built in the late sixties, which attracted sportfishermen. At about the same time, the village of Atlantic, farther down-east, also became a popular staging area for fishing on Core Banks. "A great deal of shack development (squatting) occurred during this period, mostly on Core Banks," by middle-income sportfishermen "who would simply stake out a patch, put a structure on it, and post a *no trespassing* sign on the door" (Maiolo and Tschetter, 1982:218). Local businesses responded by opening marinas, tackle shops, and small groceries.

Atlantic, according to Barbara Garrity-Blake, was first known as "Hunting Quarters." English and Scots-Irish settlers first inhabited the area. It became a distinct municipality in the late-nineteenth century and commercial fishing became its most important revenue-producing activity. By the 1930s, two of the largest seafood dealer/processing companies were located there and were owned by local residents. But not much happened in terms of demographic growth between then and 1990. During those five or six decades, the permanent population grew by 123, from 685 to 808.

Marcus Hepburn, who spent a great deal of time on Harkers Island, and Barbara Garrity-Blake, who conducted a lot of anthropological fieldwork in Atlantic and the surrounding areas, describe the villages in similar terms. They see the communities as "tightly-knit" enclaves, inhabited by creative and versatile fishermen/crofters involved in a variety of subsistence and revenue-producing activities.

Three very important infrastructure investments created the environment for change. The first was the construction of a ferry landing at Cedar Island to transport tourists to Ocracoke. Roadways were improved to handle the increased traffic, and business development responded with more shops and restaurants. Many of the tourists began to stop at the villages along the roadway, which would also include an out-of-the-way crossover to Harkers Island (Maiolo and Tschetter, 1982:218).

Second, a new bridge to Harkers Island was constructed to replace the older, wooden one that had opened on January 1, 1941, to hook up the island with the mainland. It was almost built to link up with Beaufort, which would have shortened the distance between the two communities by about eighteen miles! When I interviewed the islanders to find out why they had objected to the shorter link, they indicated that they had feared the onslaught of too many "outsiders."[6] As you can imagine, the new bridge brought more tourists anyway, and not just fishermen. Families were now drawn by the attraction of nearby Cape Lookout's pristine beaches. Several private, locally owned ferries were developed, and a family campground was opened (Maiolo and Tschetter, 1982:219).

The third development was the designation, in the late seventies, of Core and Shackleford Banks as a continuation of the National Seashore. Along with this came the condemnation of one of the motels and other property on Harkers Island in order to provide a home for the National

Park Service. The Park Service terminated the island residents' practice of building vacation huts in order to fish, garden and tend to livestock on Shackelford Banks. This met with public resistance, but since then, this restriction has become part of the accepted landscape. Some vandalism and physical encounters did occur when the Park Service Headquarters first entered the scene. The most serious incident on the island occurred in 1985, when the Cape Lookout Visitor Center was destroyed by arson. I was there at the time. It quickly became clear that this kind of behavior would not be tolerated. Even though the culprits were never caught, the decent people of the island, as they have for centuries, took steps to find ways to deal with the reality of the situation. To a large extent, residents now view the Park Service and its employees as a permanent part of the community and have moved on.[7] There remain, however, some unresolved legal issues over the use and disposal of fourteen structures under the jurisdiction of the Park Service. The winter 2003 issue North Carolina's Sea Grant publication *Coastwatch* (Green: 6–11) has an excellent historical account of the problem.

In 1982, after having time to reflect on these infrastructure changes and their immediate aftermath, Tschetter and I wrote, "Cottage development in the typical sense ..., which until recently ... [was undertaken in the Morehead City vicinity], has begun downeast mostly on Harkers Island. Property values have soared, doubling between 1978 and 1981" (Maiolo and Tschetter, 1982). And I even found some cases where prime property that had sold for a few hundred dollars in the mid-seventies had turned again for more than $20,000. We found that young adults, who were already having difficulty making a living at fishing as their parents and grandparents had, could not afford to purchase property there. We then noted that fallout from the rapid changes on and near Harkers Island promised to impact the remaining villages down-east.

As confident as I was in our predictions, what happened in the mid- and late-nineties exceeded even my expectations. The permanent population has grown to over 1,800, a sizable increase over the 1980s. But more important is the nature of the growth. The island is slowly, but steadily, in the process of converting to an upscale resort area for the wealthy. It is currently in the middle of this transition, and many traditional home sites are within view of large, luxurious, three-story vacation homes. The latter have all of the trappings you might expect—multiple decks; stacked windows, through which one can view the adjoining waterways; charming, decorative items with a nautical theme; and, of course, wet slips for recreational boats. When I examined property prices in areas with which I was most familiar, I found that even some of the *lesser* desirable waterfront lots, if there is such a thing, were in the $120,000 range.

Several interviews I conducted with young people whose families lived on the island revealed a variety of adaptive strategies for obtaining housing. When possible, a parcel or home is purchased from a relative and credit is either extended by the family or through a conventional mortgage, but this is not the typical case. More typically, a young couple, providing they choose to stay nearby, purchases an affordable parcel on the mainland, probably not near a body of water, which would make it too expensive. The couple would then either construct a home or buy a mobile home. The wife might seek a job as a waitress or a clerk. The husband might try his hand at fishing and a variety of other jobs to create his own "annual round."

Local elementary school enrollment has dropped from 300, in the late sixties, to 150, another indication of the changing demographic structure. A few of the young people go elsewhere for higher education, most to community colleges, and return for whatever professional employment is available, e.g., teaching. Others move to other, nearby areas of the coast or inland. When I approached the subject of the types of changes occurring on the island during an interview I had with an articulate lifelong resident, Karen Willis Armspacher, she sighed, "You are zeroing in on the very crisis we are facing."

Chapter Eight

It doesn't take a lot of looking to see the same trend occurring farther down-east, as far as Atlantic. "Outsiders" are beginning to buy up property at prices "the likes of which we ain't seen," according to one of my respondents. While the locals think the values are outrageously high, the buyers, who have meticulously compared prices along the coast in three states, find them to be a "steal." After the purchase, they often tear down the existing buildings and rebuild in the same manner I described above.

Thus, just as we saw to the north, the kinds of development during recent decades may be economically beneficial to the region and the state, and they certainly bring a great deal of recreational pleasure to those who vacation in or move into the central coastal area. All of this is happening at about the same time the commercial fishing industry, in general, and the shrimp fishery, in particular, are under stress and may be on the decline. Garrity-Blake conducted interviews with traditional fishermen who saw the handwriting on the wall regarding what they saw as the increasing influence of recreational fishing and housing development in the down-east region. One fisherman complained, "Sportsfishermens gonna stop this business ... they got too much control ...." From a fish-factory foreman's wife, she heard, "It makes me sick to my stomach what they're building now .... Everything can't be residential, who's doing the working? ....outside money comes in, outside money comes out, right back to Raleigh. They don't contribute a thing to the economy of Carteret County.... We'll destroy ourselves."[8]

Many times these kinds of complaints, as I mentioned before, come from people who have some appreciation for the economic benefits that derive from the influx of vacationers. But they are torn by the fact that they see their way of life threatened and often reflect upon it with a disarming honesty.

One person I know very well has converted his annual round from exclusively commercial fishing, and preparations for it, to part-time fishing and running a very successful business. His fishing has been changed to exclude channel-netting in favor of concentrating on trawling for crabs and kicking for clams. I cannot reveal the nature of his current business, because it is so unique that it might give him away. He is a shy person, and I want to protect his anonymity. When I asked him why he had changed his fishing style, his answer was, "I can't make a livin' at it no more. Besides, ya know, I'm gittin' up there [in age]." On the subject of vacation housing development in his beloved village: "I wished things would be the same, but they ain't and that's it. But the dingbatters [vacationers] give me a heckuva livin' at the other thing, ya know. So things have a way of workin' out. I just wished the kids could have some of it the way we usta. But they won't, and that's that. They go on off to workin' in Cherry Point and Morehead and that's okay. So it all works out for the best, I reckon. But I sure wished the kids could have some of it like we usta."

A fish dealer in the same community noted, "Man the ssrimp ain't worth much a damn no more, but if I can git some stripers, I can make a killin'. Clams can be good, but, oh, man, those crabs is gone. Can't make a livin' at that no more." On the subject of vacationers and their expensive homes, he said, "It sure has changed around here. My kids come back and don't even rekanize it here. I don't know much how I feel about it, there's some good and some bad." A clerk in a large store remarked, "We can't afford to live here. I come to work cause I like it here, but Lord knows, we can't afford no property here. My husband feeshes here, too, but we can't afford it, no way, to buy here. I can make a decent livin', though, with this job and all."

On Harkers Island, an interesting adaptation, for some, is the expansion of the cottage industry of making and selling duck decoys, an age-old tradition of craftsmanship on the island. The finished products are magnificent and make wonderful decorative souvenirs. The craft has become such a serious business that the Core Sound Carver's Guild now has its own building on the road leading to Harkers Island. There the artisans have space to ply their craft and learn from each other. An annual event is held for the public to visit the guild to admire and purchase the decoys.

Chapter Eight

The centerpiece showcasing the island's history, culture, and diversification is the brand-new Core Sound Waterfowl Museum, which also showcases the best of the decoys. My friend, Professor Jim Sabella, of the University of North Carolina at Wilmington and an expert on the island's history and culture, calls the new facility "exquisite." He has noted that it serves the dual use of being both a repository for the island's history and a wonderful place for tourists to visit. And there has to be a place in Heaven for all of the hard work and dedication of Karen Willis Amspacher, who transformed the museum from a tiny building on the island to a masterful facility that can accommodate tourists, workshops, and conferences.

The South Coast As we move south along the coast, the changes are, again, familiar. The characteristics of the growth, however, differ from that in the other two areas. First, the area is dominated by Wilmington, the city with the largest port in the state and the largest population of any coastal town. But the city does not have a history of extensive influence by commercial fishing. Situated in New Hanover County, Wilmington extends its influence north to Pender County and south through Brunswick County to the South Carolina state line and the Myrtle Beach resort area. I will focus on the area from New Hanover to the south, where virtually all of the resort development has taken place.

The Cape Fear River attracted the original settlers to the Wilmington area. The city is thirty miles upriver from Southport and the ocean, which is the main reason it never developed as a center for commercial fishing harvesting activity, although it did serve, for a while, as an important fish-product distribution center. Early-seventeenth-century attempts at colonization failed, and Wilmington was not incorporated until 1740. Corn and rice were grown for local consumption, and beef and pork were harvested from free-range cattle and swine. But the city flourished as a port and a center for shipbuilding. Products such as tar, pine pitch, and turpentine were important parts of its export economy. In fact, there was a time when the area was considered to be the naval-stores capital of the world. This promoted the development of slave-based plantations specifically designed to produce those goods for international export. Later, for a while, the city was home to the largest cotton exchange in the world (Lee, 1971; Daniel and Moore, 2002). The old exchange, situated on the restored Wilmington waterfront, is now filled with shops and restaurants.

After the Civil War, the city continued to export naval stores, along with rice, cotton, peanuts, and lumber. Freedom from slavery allowed African-Americans to participate in the flourishing economy. To the south they participated in the fisheries. In the city of Wilmington, a black middle class developed, and the state's first black lawyer, Mr. George Lawrence Mabson, and first black physician, Dr. James Shober, resided there.

As I mentioned earlier, shipbuilding is an important part of Wilmington's history, beginning with its first colonization and the creation of the lumber industry. With each of the world wars, the industry was revitalized, and hundreds of vessels were constructed during those conflicts (more than 200 in the 1940s alone). During the same period, the North Carolina legislature created the State Port Authority. This assured that the boat-building infrastructure could be transformed into a permanent, economically viable industry.

In addition to Wilmington area's water transportation system, for much of its late-nineteenth- and twentieth-century history, railroads played a major part in the local economy. Since the 1970s the city has undergone a transformation that includes restoration of its charming historical district and the successful recruitment of major industries (e.g., Corning Glass, General Electric, pharmaceutical companies, and a film production studio). It also now boasts a major convention center, an active cultural arts program, and a robust nightlife. It is the educational hub of the south coast, as well. The nearby islands of Wrightsville, Carolina Beach, and Kure Beach have been, and continue to be, an important part of the economy since the late nineteenth century, when they began serving as family vacation resorts.

To the south, in Brunswick County, lies Southport, the birthplace of the North Carolina shrimping industry. A community of fishermen, merchants, and boat captains grew around Fort Johnson, which was built in 1754. First established as Smithville, Southport's original name, it became the county seat in 1808. As an important commercial activity, fishing was late coming to Southport and Brunswick County (Carson, 1992; Daniel and Moore, 2002; Lee, 1980). It is, at best, of modest economic significance in Southport today. Shrimping remains an important activity for fewer than two-dozen boats in Varnamtown to the south.

Wilmington served as the *market* hub for the entire south region of the state. But until refrigeration came in the 1870s, its function in *seafood* marketing was limited to handling salted fish and some shellfish. It wasn't until 1879 that the region entered into the commercial fishing industry when the first commercial fishing vessel went to sea out of what is now Southport (Lee, 1980:221). Shrimp catches were not even reported until 1897. Commercial food fishing became an integral part of the economy beginning in 1902. There was an attempt to establish a menhaden fishery in the area in 1871. It failed, but was later revived in Brunswick County soon after World War I. It survived until the mid-eighties, when South Carolina implemented regulations that prohibit trawling within three miles of the state's shoreline. The region's menhaden processors and fishermen had relied heavily on the stocks south of border.

The permanent population of New Hanover doubled between 1970 and 2000, with a 33% increase occurring in the nineties alone. Brunswick nearly tripled its population during that thirty-year period, including an increase of 44% in the nineties. Two major kinds of development occurred along Route 17 and on the nearby islands. As was the case along the entire coastline, they experienced a boom in vacation villas and homes, along with hotels and motels on the islands of Wrightsville, Carolina, and

Kure, as well as on Holden, Ocean Isle, and Sunset Beaches. The number of vacation units for short-term rentals, on the comparatively small island communities in New Hanover County rose to over 4,000. But the larger islands to the south in Brunswick saw an increase to about 15,500 rental units during the same period. I will come back to reexamine this issue further at the end of this chapter.

There was also a tremendous influx of people moving into the area to service the expanding commercial port, the military port near Southport, the University of North Carolina at Wilmington, the tourist industry, and the burgeoning recreational industry. In addition to recreational fishing and boating, golf became a major part of the local economy.

During the seventies and eighties, Myrtle Beach, South Carolina, became somewhat of a Mecca for fall and winter golfing and summer beach resort activities. This industry spilled over to the north as property became more expensive in South Carolina and North Carolina discovered the value of "golf packages." That concept became so popular among Canadians, as well as other "snowbirds" from northern states, that about one-half of Myrtle Beach's economy became dependent upon it.

The influence of Myrtle Beach's golfing industry migrated to the north and penetrated Brunswick County with the development of a quite impressive restaurant trade in Calabash. The village became famous for its southern-fried seafood, served in enormous quantities for a fairly low price. The intent was to lure golfers vacationing in the Myrtle Beach and surrounding areas to drive to Calabash for a southern-fried seafood meal. At last count, there were sixteen restaurants, most of them serving seafood, in a town of 711 people! These establishments are designed to move a lot of people in and out quickly. The concept became so popular among golfers and vacationing families that it was transported south into Myrtle Beach itself.

Next came the development of golf courses inside North Carolina along the state line with South Carolina. While golf-course development

Chapter Eight

also occurred in the other two coastal regions we discussed, it pales in comparison to what happened in New Hanover and New Brunswick counties, both in terms of numbers and, in golf parlance, "the quality of the tracks." That is a reference to design and playing features. Golfers know!

This industry now extends from about twenty miles north of Wilmington, beginning at Topsail Greens and then Belvedere Plantation, south to Carolina Shores in Calabash. In between those points lie some of the most up-to-date and varied golf links, catering to every level of skill—and price! Most of these are tied into a community setting that caters to retirees from outside of the region and golf-package patrons. The most *expensive* of these (which does not cater to visitors) is Landfall, located between Wilmington and Wrightsville Beach. It is affordable only for the very wealthy. The *largest,* St. James Plantation, west of Southport, has been planned ultimately to have thousands of homes, villas, timeshares, and four golf courses. At last count, the development already had twenty-two hundred properties sold, seven hundred homes built, another one hundred under construction, three eighteen-hole golf courses, and one nine-hole course.

Let's now contrast this recreational development with the contemporaneous changes in commercial fishing. During eighties, the number of full-time commercial fishermen in the two counties increased about 27%, from a combined total of 550 to 700. Much of this was due to the gentrification of commercial fishing that I discussed earlier. Rumors of the impending moratorium caused a spike in license applications during the '93/'94 season. The number of part-time commercial fishing-license applications remained steady during the nineties and did not spike when word got out about the moratorium.

What is very interesting about this period is the number of active dealer licenses: 96 in Brunswick County and 63 in New Hanover County. Yet the local landings are comparatively low, especially in New Hanover. For example, a revisit to the table in Chapter Seven will reveal that 90 dealers handle 31.6 million pounds of seafood in Dare County, and 125 operators produce 14 million pounds in Carteret County. Brunswick County's 96 dealers handle only 2.2 million pounds, and 63 dealers in New Hanover County handle only about 1.7 million. With the exception of those who handle offshore finfish catches, most of these are comparatively small dealers who are more oriented toward handling seafood for the local markets than in the other two regions I have discussed. And remember my reference in Chapter Two to the sixty-two trawlers in that community in 1932 and my references to the community as the birthplace of shrimping in North Carolina. In February 2003, there were two local shrimp trawlers in Southport (along with six snapper/grouper boats), and fewer than two-dozen in nearby communities! The largest employer in the county was the school system, with 1,800 employees. The local community college employs 300 people, the county government 827, another 291 work in community governments, and 350 in the hospital, bringing the governmental, educational and health service workforce to nearly 3,600 people. Dupont is the largest private employer (1,100), followed by the local nuclear plant (980). Companies that handle and process seafood employ only about 100 people. That number changes somewhat during the height of the shrimping season in Varnamtown, when some local women and children are still involved in heading shrimp.

    I would suggest that the data from the south coast, especially from New Hanover County, indicate that the yearly round of activities within the communities there would be quite different from the other two areas discussed. Golf is a recreational activity, but it is year round in that region. Add this to the in-migrants who are still in the workforce and the picture is quite different from the central and northern coasts. And anyone who has experienced Wilmington's year-round traffic jams will support this assertion.

LOOKING INTO THE FUTURE Among the many interesting opportunities I had to do coastal-zone and fisheries research during my career at East

Carolina University, one of the more enlightening was to collaborate with Paul Tschetter to study the impact of population growth on pollution and, therefore, certain shellfish-harvesting activities. In addition to producing several publications on the subject, we did a report for Sea Grant that contained an interesting statistical analysis (Maiolo and Tschetter, Summer 1983). Tschetter showed me how to separate recreational populations analytically into two basic categories. The first is the day-recreation group, consisting of those who travel to the coast for a one-day fishing trip or to visit the beach. The second is what he termed the "overnight" population. These are the visitors who either own or rent space for the range of their vacations—from one night to an entire season. He made the argument that the latter was the more important of the two statistics. The overnighters are the people who contribute to the local economy through the creation of construction and service jobs, and, unlike the day-recreation group, place demands on housing facilities.

By dividing the Census Bureau's total number *housing units* by the number of *households* it counts, one can estimate a given locality's "recreational-housing capacity" (i.e., non-hotel/motel housing available for vacation rentals). This also serves as a rough indication of the relative importance of recreational housing to the community's economy. This might seem confusing, but a simple example will make it clear. If there are six housing units and four households, then four out of every six housing units are occupied by a family on a permanent basis. Dividing 6 by 4 yields a ratio of 1.5:1. *Fewer* permanent households (e.g., three) would produce a *greater* ratio (in this case, 2.0:1), which indicates that *more* units are available for rental. (1.0 is the smallest possible ratio, indicating that all housing units are permanently occupied.) After having completed this task, one can then compare growth over time *among* various communities, and then look at that growth in comparison to *overall* population growth.

Among the nine counties we studied, we examined data from the five counties we have discussed in this chapter—Dare, Hyde, Carteret, New

Hanover, and Brunswick—over the thirty-year period from1950 to 1980. I exclude data from Hyde County, however, because it is too difficult to sort out the changes there, given the small sizes of its communities, including Ocracoke.

The largest ratio Tschetter and I found to have occurred during the period was in Dare County, followed, in decreasing order, by Brunswick, Carteret, and New Hanover. In absolute, non-relative terms, we found Carteret County to have had the largest *net* increase in recreational-housing capacity during the period. I repeated the procedure for the year 2000 and found precisely the same order. The differences I found involved the magnitude of the changes in the ratio, as illustrated in Table 8.1. The largest *change* in ratio occurred in Carteret County, a rather astounding 25% increase in vacation housing in relation to total housing. Brunswick was second in line, followed by Dare. New Hanover actually experienced a decline in its ratio.

When one compares the changes, here is what I believe the data suggest. Data from Carteret reveal that vacation-home development has significantly outstripped housing for permanent residents, a population that grew by 44% during the last twenty years of the century. And while most of this has occurred on Bogue Banks, the down-east villages are feeling the changes as well.

TABLE 8.1 Estimated Recreational-Housing Capacity as Implied by Ratios of Total Non-Hotel/Motel Housing Units to Permanent Households

County	Ratio 1950–1980	Ratio 1980–2000	% Change: Ratio 1980–2000	% Change: Population 1980–2000
Dare	1.90:1	2.10:1	+10.5	124
Brunswick	1.48:1	1.69:1	+14.1	116
Carteret	1.29:1	1.62:1	+25.6	44
New Hanover	1.17:1	1.15:1	-0.17	58

I was able to obtain specific information about Harkers Island for the winter of 2002/03, and compare the data with estimates I made in the late-eighties. Recreational-housing capacity had grown to 1.71 from between 1.05 and 1.10, or more than a 50% increase. During the same period, the population grew by about 20%. And, as I noted, the number of children enrolled in the elementary school declined from about 300 to 150.

In January 2003, the total number of houses on the island amounted to about 1,200, of which 500 served as vacation homes. Some have been built to accommodate future retirements. These have been constructed for what one of my respondents called "pre-retirees," who build homes and use them for vacations with the intention of making them their permanent residences upon retirement. The respondent also indicated that re-sales of these homes often occur because one of the spouses becomes ill and/or dies. When this happens, the survivor frequently chooses to move (or stay) closer to the couple's children and grandchildren. She complained that, at least when she was growing up, the few people who owned vacation homes on the island were "permanent" tourists. They would return every year, appreciated the local customs, and their children became lasting friends with the year-round residents.

Housing available as vacation rentals grew by 14% in Brunswick during the period under consideration, but the population grew by an even more impressive 116%. This is in line with my earlier discussion, you may recall, which focused on the development of retirement communities there. And the fact that New Hanover had a *negative* growth in recreational housing indicates two things: first, the county's islands had reached a saturation point in the eighties, partially due to successful political action to restrict further growth, particularly on Wrightsville Beach, where development had outstripped the area's "carrying capacity" (see Maiolo and Tschetter, 1981); and second, a great deal of in-migration occurred to occupy permanent housing, a significant amount of which was accounted for by retirement couples moving into upscale developments like Porter's Neck and Landfall during the nineties.

Dare County appears to have had it both ways, as well. Its population grew the most among the coastal areas we studied and also experienced more than a 10% growth in vacation rentals. What this suggests to me is that its appeal as a vacation destination did not suffer a bit as more and more people found the region attractive as a location for permanent residence.

As part of our study that found increases in recreational housing, we also examined changes in the occupational profiles of the counties. For the coast overall, we observed a significant relative *decrease in proportionate dependence* on primary extractive industries, of which fishing was the most dominant (Maiolo and Tschetter, 1981:11–14). Yet we found overall employment in all occupational categories (combined) had increased. We concluded, "the development of recreational activity was a primary stimulus for growth" across all of the study counties.

A recent study by three economists, which focused on the impact of recreational fishing on the state's economy, supports our thesis. Steinbeck, Gentner, and Castle (2003) concluded that the annual *direct* economic impact of recreational angling in North Carolina is over $600 million. Indirect impacts (through other goods and services) total $139.5 million, and its contribution to wages and salaries is another $198 million through the creation of over 15,000 jobs.[9] According to Gentner, just under one-third of these monies is contributed by out-of-state anglers who come to North Carolina and pay for lodging, fuel, bait, charters, transportation, and so on (personal communication with the author, 2003).

It is difficult not to feel some concern about these trends, which are, in some ways, inimical to commercial fishing, at least as we have known it in the past. On the other hand, even though the industry is losing ground in *relative* terms, the number of people out there who still take the chance that putting a net in the water will yield a decent and rewarding living has stayed steady in *absolute* terms during the past decade. What I mean is this: in 2002, the number of fishermen operating under the system that was put in place in 1997 was about the same in all three regions as it was in 1994, just before the moratorium became effective. This seems also to be true in the shrimp fishery, although, as we have seen, many of these fishermen have changed their emphases and annual rounds to include land-based jobs. As I noted in Chapter Six, this may be interpreted as an adaptive strategy to continue with a hybrid annual round that includes non-fishing activities, such as a land-based job. Recall that we found this pattern already present when we did our original fieldwork in the eighties. Garrity-Blake (1996:7) found such hybrid annual rounds to be more common in urbanized coastal areas.

The State of North Carolina suffered through the aftermath of several hurricanes during the nineties and then found a way to rebound, especially following the devastating impact of Hurricane Floyd in 1999 (see Maiolo, et al., 2001). Hurricane Isabel's visit to the coast in September 2003 left many of the central and northern waterside communities devastated once again, including more than a few of those I have described here—Harkers Island, Atlantic, and other down-east villages, as well as Ocracoke, Hatteras, and communities throughout Hyde County. It will take time, but North Carolina will slowly recover economically, providing a concomitant economic recovery also occurs at the national level from the high-tech "bust" at the turn of the century. The communities on the coast have a remarkable history of adaptation, as do the fishermen and their families.[10] Now we are experiencing an incredible surge in the number and diversity of the people who want to experience that history and help craft its future. Depending upon whom you ask, the coast is either becoming a treasure trove of cultural diversity or an ecological and cultural disaster. No matter what, though, the shape of the *future* is in the hands of those who are now experiencing the *present*. I can only hope that, as in the past, diversity is seen as a strength rather than as something to fear and that everyone who wishes to enjoy the coast shares the same goal—the preservation of our precious resources. And I hope, as well, that commercial fishing in North Carolina will continue to be seen as one of those resources.

Endnotes

PREFACE

1. Our wide-ranging and in-depth work from the 1980s in this area, along with our treatments of the development of harvesting technology and other crucial issues in the shrimp fishery, were first utilized and then obliquely referred to in a recent book by Kelly and Kelly (1993) as coming from an "unsigned report" (p. 27).

2. At about the same time, James Easley undertook important economic research at North Carolina State University under the sponsorship of the N.C. Division of Marine Fisheries.

ACKNOWLEDGEMENTS

1. The original title I chose back in the 1980s was: *Nickel A Bucket: The History and Culture of the North Carolina Shrimp Industry.*

CHAPTER TWO

1. William Still contributed a great deal of effort to the historical analysis of and the narrative describing the fishery, which served as an important basis for the report we completed in 1981 and an article he published in 1987. For this reason, he is deservedly designated as co-author of this chapter and Chapter Four.

CHAPTER THREE

1. My colleague Marcus Hepburn contributed a great deal to the early fieldwork for this chapter and contributed greatly to my understanding of the channel-net fishery.

2. In recent years, the use of this technique has evolved and became firmly embedded in the shrimp fishery in the southern part of the state. I will discuss this throughout the chapter.

3. Another local innovation is shell-stocking and machine processing of scallops—a topic I discuss briefly later in the book. There is also some evidence that a few shrimpers in

Texas experimented with channel nets. It is not known whether this was due to the adoption of the North Carolina technique or an independent development.

CHAPTER FOUR

1. As I note in Chapter Two and elsewhere, William Still conducted a great deal of the archival research that served as the basis for a report we co-authored on some aspects of the shrimp fishery's development. Much of this chapter is based on that report (Maiolo and Still, 1981). As I note in the Preface, when it became clear the book project would be delayed, Bill published some of the materials in *American Neptune* (1987).

CHAPTER FIVE

1. I am grateful to William Still who assisted in gathering the materials related to the historical development of the agency, which eventually became the North Carolina Division of Marine Fisheries and the Marine Fisheries Commission. His work served as the basis for this section of the chapter.

2. There were several types of endorsement to sell; including one for the vessel, one for the sale of aquaculture harvests, one the sale of some fish caught in fishing tournaments, and yet another for non-residents. In order to keep the discussion as simple as possible, I will limit it, at this point, to the vessel endorsement.

CHAPTER SIX

1. This chapter is based upon my own fieldwork, which began in the 1980s, was interrupted during the early 1990s by other professional duties, and was resumed during the 1990s. My use of the terms "Commercial Man" and "fishermen" is not intended to show disrespect for the many women who are commercial fishers. In my fieldwork, though, I have not yet met a female shrimper, even though I know there have been females in the fishery. For this book, then, the traditional terms seem to work.

2. Each year's adjusted or deflated value (AV) was computed according the following formula: AV=AYV x 100, divided by YCPI, where AYV is the actual year value And YCPI is a given year's CPI in relation to 1967. For example, the 1970 CPI is 116.3, indicating a 16.3% increase in the cost of living from 1967 to 1970.

3. For information on the fluctuations in the state's entire fishing industry, the reader is encouraged to visit the North Carolina Division of Marine Fisheries website and browse through the statistics section.

4. It is interesting to note that neither our fieldwork in the eighties, mine later in the nineties and then in 2003, nor Brian Cheuvront's 2002 work found a single female in the shrimp fishery. He did find, however, six females out of the total of 259 Core Sound commercial fishermen he studied, and three female shrimpers in his study of the remaining coast (2003).

5. These are eighth-grade graduates. I am grateful to Karen Willis Amspacher and her brother Mark Willis for obtaining this information. I will say more about the school enrollment in Chapter Eight.

6. For example, the Saltonstall-Kenndy Act. Administered by NMFS, this legislation has provided financial assistance for research and development projects to benefit the U.S. fishing industry since 1954.

CHAPTER 7

1. The reason for the title of this chapter will be made clear later on in the chapter. I think you will find it justified. My colleague, John Bort, framed much of the fieldwork he and I conducted in the 1980s that served as the basis for both a report we did together and this chapter. Furthermore, he took the lead in our writing several of the sections of reports that resulted from that fieldwork. I am grateful for his excellent, significant contributions. The fieldwork on seafood dealers after that period is solely my responsibility.

2. John Bort deserves a great deal of credit for educating me on the intricacies of adaptive strategies.

3. In addition to loans provided through the Saltonstall-Kennedy Act, which are mainly pursued by large-scale operators, small-scale fishermen can obtain assistance through the U.S. Small Business Administration.

4. A critical examination of the seafood-processing labor force in North Carolina can be found in David Griffith's *Jones's Minimal: Low Wage Labor in the United States* (Albany: State University of New York Press, 1993).

5. An evisceration machine was developed in North Carolina (by whom is debatable) and was originally intended for removing the shells from shrimp. That did not work. However, the concept was re-engineered to separate the adductor muscle of the scallop, the part that is eaten, from the rest of the animal. The machine consists of a number of rollers that, by turning and re-turning, pinch the muscle clean.

6. Curly informed me that his company began seeking and purchasing shrimp imported from Ecuador in the late-seventies and early-eighties, when North Carolina's shrimp harvest was "goin' up and down like a rollah coasta."

CHAPTER EIGHT

1. The islands that extend to the west and south of Morehead City are also known as Barrier Islands, but are not part of the Outer Banks. The latter are more distant from the mainland, whereas the former are as close as one-half mile.

2. A great deal of the narrative in this and the next section is based on two projects for which I was privileged to serve as principal investigator. The first was funded by Minerals Management service. My administrative partner was John Petterson of Social Impact Assessment, Inc. We had a wonderful staff of young people who did the bulk of the on-the-ground fieldwork and provided high-quality data and analysis. I am particularly grateful to the researchers on my staff, Robert and Belinda Blinkoff, Ed Glazer, and the always-dependable Barbara Garrity-Blake. The second was funded by the Center for Transportation and the Environment and part of this research was aided, again, by John Petterson and his fine staff. A portion of this project was also improved by the involvement of Jeff Johnson.

3. The county was named in honor of Virginia Dare, believed to have been the first child born of English parents in America.

4. In an earlier presentation (May 1996) I referred to such annual changes as "fission and fusion" in coastal communities.

5. Dare County does not include Ocracoke, which is in Hyde County. This is of great concern to many in Ocracoke, who believe that culturally and economically, its security is more closely tied to Dare County.

6. Carmine Prioli (1998:24–26) has a slightly different take on his.

7. Prioli (1998:35–37) presents an informative narrative on the comparison of the designation of Core vs. Shackleford Banks and how the two differed in regard to their use before this designation. His treatment is another reminder of the thesis Ellis presented (1986) that I discussed earlier in this chapter—i.e., that communities might seem superficially similar, but differ greatly beneath their surfaces. His narrative on the cooperative effort between the local residents and the federal government "to bridge the gulf" and preserve the local heritage is also well worth reading (p. 39).

8. These comments were part of Garrity-Blake's section of the September 30, 1993, report to the Minerals Management Service (pp. 264–265).

9. There is another $959 million in impacts *in other states* from spending by anglers who fish in North Carolina. Thus, the total impact exceeds the astounding figure of $1.5 billion!

10. I strongly recommend reading David Stick's books, in this regard, especially *Outer Banks of North Carolina* and *Graveyard of the Atlantic*, both of which are published by the University of North Carolina Press. These books chronicle the lifestyles and survival of coastal residents and commercial fishermen under incredibly difficult historical circumstances. I also recommend Lee Maril's 1986 book about changes along the Texas Gulf Coast. His work chronicles patterns similar to those I have described here.

Bibliography and References

Acheson, J.M. "The Lobster Fiefs: Economic and Ecological Effects of Territoriality in the Maine Lobster Fishery." *Human Ecology* 3 (1975): 183–207.

Andersen, R. "Public and Private Access Management in Newfoundland Fishing." In *North Atlantic Maritime Cultures*, edited by R. Andersen. The Hague: Mouton, 1979.

Atlantic States Marine Fisheries Commission (ASMFC). "Southern Shrimp." *Marine Resources of the Atlantic Coast* Leaflet Number 4. Tallahassee, Florida: Atlantic States Marine Fisheries Commission, 1965.

———. *Estimates of Finfish Bycatch in the South Atlantic Shrimp Fishery*. Final report of the Southeast Area Monitoring and Assessment Program, South Atlantic Committee, Shrimp Bycatch Workshop. Washington, D.C: Atlantic States Marine Fisheries Commission, n.d.

Blomo, V., Michael Orbach, and John Maiolo. "Competition and Conflict in the Atlantic Menhaden Industry." *American Journal of Economics and Sociology* 47 (1988):41–60.

Bort, John, and J. Maiolo. "The Marketing of Shrimp." In *The Sociocultural Context and Occupational and Marketing Structures of the North Carolina Shrimp Fishery*, Third Year Report, Vol. 2. Greenville, North Carolina: East Carolina University, Department of Sociology and Anthropology, 1980.

Calder, D.R., P.J. Eldridge, and M.H. Shealy, Jr. "Description of the Resource." In *The Shrimp Fishery of the South Atlantic United States: A Management Planning Profile*, South Carolina Marine Resources Center Report Number 5, edited by Dale Calder, et al. Charleston: South Carolina Marine Resources Center, 1974.

Captiva, F.J. "Trends in Shrimp Trawler Design and Construction the Past Five Years." In *Proceedings of the 19th Annual Session of the Gulf and Caribbean Fisheries Institute*. Charleston: Gulf and Caribbean Fisheries Institute, 1967.

Carson, Susan. *Joshua's Dream: The Story of Old Southport A Town With Two Names*. Southport, North Carolina: Southport Historical Society, 1992.

Chestnut, A.F., and H.S. Davis. *Synopsis of Marine Fisheries*. University of North Carolina Sea Grant Program Publication UNC-SG-75-12. Raleigh: University of North Carolina Sea Grant Program, 1975.

Cheuvront, Brian. *A Social and Economic Analysis of Commercial Fisheries of Core Sound, North Carolina*. Morehead City: North Carolina Division of Marine Fisheries, 2002.

Daniel, Deborah, and Gwyne Moore. *Insider's Guide: North Carolina's Southern Coast and Wilmington.* 9th ed. Wilmington, North Carolina: Sea Publications, Inc., 2002.

Earll, R.E. "North Carolina and Its Fisheries: Part XII." In *The Fishery and Fishery Industry of the United States, Section II.* Washington, D.C.: G.B. Goode and Staff, 1887.

Eldridge, P.J., and S.A. Goldstein, eds. *The Shrimp Fishery of the South Atlantic United States: A Regional Management Plan,* South Carolina Marine Resources Center Technical Report No. 8. Charleston: South Carolina Marine Resources Center, 1975.

Ellis, Carolyn. *Fisher Folk: Two Communities on Chesapeake Bay.* Lexington: University Press of Kentucky, 1986.

Farfante, P. "Western Atlantic Shrimps of the Genus *Penaeus.*" *Fishery Bulletin* 67, no. 3 (1969): 461 passim.

Fisheries Moratorium Steering Committee. *Final Report to the Joint Legislative Commission on Seafood and Aquaculture of the North Carolina General Assembly.* North Carolina Sea Grant College Program. UNC-SC-96-11. Raleigh: North Carolina Sea Grant College Program, October 1996.

Garrity-Blake, Barbara. *The Fish Factory.* Knoxville: University of Tennessee Press, 1994.

———. *To Fish or Not To Fish: Occupational Transitions Within the Commercial Fishing Community, Carteret County, North Carolina.* Raleigh: University of North Carolina Sea Grant, 1996.

Green, Ann. "Core Banks Cottages Rich in History, Tradition." *Coastwatch* (Winter 2003): 6–11.

Griffith, D., and J. Maiolo. "Considering the Source: Testimony vs. Data in Controversies Surrounding Gulf and South Atlantic Trap Fisheries." *City and Society* 3, no. 1 (1989): 74–88.

Griffith, David. *Jones's Minimal: Low Wage Labor in the United States.* Albany: State University of New York Press, 1993.

———. *Characterization of the North Carolina Recreational Shrimp Trawl Fishery: A Preliminary Analysis.* Raleigh: University of North Carolina Sea Grant College Program, 1996a.

———. *Impacts of New Regulations on North Carolina Fishermen: A Classificatory Analysis.* Raleigh North Carolina: University of North Carolina Sea Grant College Program. 1996b.

———. *The Estuary's Gift.* University Park: Pennsylvania State University Press, 1999.

Imperial, Mark, and Tracy Yandle. "Taking Institutions Seriously: Using the IAD Framework to Analyze Fisheries Policy." Paper presented at the spring meeting of the Southern Division, American Fisheries Society, Wilmington, North Carolina, February 14–16, 2003.

Jensen, A.C. *A Brief History of the New England Offshore Fisheries.* U.S. Dept. of the Interior, Bureau of Commercial Fisheries. Leaflet 594. Washington, D.C.: U.S. Dept. of the Interior, Bureau of Commercial Fisheries, 1967.

Johnson, F.F., and M.L. Linder. "The Shrimp Industry of the South Atlantic, With Notes on Other Domestic and Foreign Areas." In *Investigational Reports of the U.S. Bureau of Fisheries,* Vol. 1. Washington, D.C.: U.S. Department of the Interior, Bureau of Commercial Fisheries, 1934.

Johnson, J.C., and Michael Orbach. "Migratory Fishermen: A Case Study in Interjurisdictional Natural Resource Management." *Ocean and Shoreline Management* 13 (1990): 231–252.

———. *Effort Management in North Carolina Fisheries: A Total Systems Approach.* Raleigh: University of North Carolina Sea Grant College Program, 1996.

Joyce, E.A., Jr., and B. Eldred. *The Florida Shrimping Industry.* Tallahassee: Florida Board of Conservation, 1968.

Kelly, R., and B. Kelly. *The Carolina Watermen.* Winston-Salem, North Carolina: John F. Blair Publishers, 1993.

Knopf, G.M. "Opportunities in the Shrimp Fishery Industry of the Southeastern United States." *Sea Grant Information Bulletin* 3 (January 1970): 16.

Lee, Lawrence. *The History of Brunswick County North Carolina.* Charlotte North Carolina: Heritage Press, 1980.

———. *New Hanover County ... A Brief History.* Raleigh North Carolina: Division of Archives and History, Department of Cultural Resources, 1984.

Lemon, J.M. "Developments in Refrigeration of Fish in the United States." In *Investigational Reports of the U.S. Bureau of Fisheries.* Washington, D.C.: U.S. Bureau of Fisheries, 1936.

Maiolo, J. *Social, Cultural and Economic Dependency on Fishery Resources in the Study Area, Technical Report Number 1.* Prepared for the Center for Transportation and the Environment, University of North Carolina Institute for Transportation Research and Education. North Carolina State University, Raleigh. East Carolina University: Institute for Coastal and Marine Resources, 1994.

———. *Using Socioeconomic Information to Assess the Impacts of Surface Transportation in North Carolina Coastal Communities, Technical Report Number 3, Survey Results.* Prepared for the Center for Transportation and the Environment, University of North Carolina Institute for Transportation Research and Education. North Carolina State University, Raleigh. Greenville, North Carolina: Institute for Coastal and Marine Resources, East Carolina University, 1995.

Maiolo, J., Albert Delia, John Whitehead, Bob Edwards, Kenneth Wilson, Marika Van Willigen, Claudia Williams, and Melanie Meekins. *A Socioeconomic Hurricane Evacuation Impact Analysis for Coastal North Carolina: A Case Study of Hurricane Bonnie.* Report to the North Carolina Department of Crime control and Public Safety. Greenville, North Carolina: Regional Development Services, East Carolina University, 1999.

Maiolo, J.R. *Implications of Proposed Management Measures on N.C. Sea Scallop Fishermen and Dealers.* Raleigh: University of North Carolina Sea Grant, 1981.

Maiolo, J.R., Edward Glazier, Miriam Young, John Petterson, and Michael Downs. *Using Socioeconomic Information to Assess The Impacts of Surface Transportation in North Carolina Coastal Communities, Technical Report Number 2, Community Studies.* Prepared for the Center for Transportation and the Environment, University of North Carolina Institute for Transportation Research and Education, North Carolina State University. Greenville, North Carolina, and LaJolla, California: Institute for Coastal and Marine Resources, East Carolina University, and Impact Assessment, Inc., 1995.

Maiolo, J.R., and M.K. Orbach, editors. *Modernization and Marine Fisheries Policy.* Ann Arbor:Ann Arbor Science Press, 1982.

Maiolo, J.R., and J. Petterson, with E. Glazier, B. Blinkoff, R. Blinkoff, B. Garrity-Blake, M. Downs, H. Edwards, J. Bourne, M. Roden, L. Schernthanner, D. Lavin, P. Dziuban, and J.Wahlstrand. *Community Studies,* Volume 3 of the *Final Report for the Coastal North Carolina Socioeconomic Study.* Submitted to the U.S. Dept. of Interior, Minerals Management Service. Greenville, North Carolina, and LaJolla, California: East Carolina University and Social Impact Assessment, Inc., 1993.

Maiolo, J.R., and William Still. "Historical Development of the Shrimp Fishery." In *The Sociocultural Context and Occupational and Marketing Structures of the North Carolina Shrimp Fishery,* Second Year Report, Volume 1. Greenville North Carolina: East Carolina University, Department of Sociology and Anthropology, 1981.

Maiolo, J.R. and Paul Tschetter. "Infrastructure Investments in Coastal Communities: A Neglected Issue in Studies of Maritime Adaptations." In *Modernization and Marine Fisheries Policy,* edited by J.R. Maiolo and M.K. Orbach. Ann Arbor: Ann Arbor Science (Butterworth Group), 1982.

Maiolo, J.R., and P. Tschetter. "Relating Population Growth to Shellfish Bed Closures: A Case Study from North Carolina." *Coastal Zone Management Journal* 9 (1981): 1.

Maiolo, J.R., John C. Whitehead, Monica McGee, Lauriston King, Jeffrey Johnson, and Harold Stone. *Facing Our Future: Hurricane Floyd and Recovery in the Coastal Plain.* Wilmington, North Carolina: Coastal Carolina Press, 2001.

Maiolo, John R. "Fission and Fusion in Coastal Communities: Implications for Social Policy Initiatives." Paper presented at the Sixth Annual International Symposium on Society and Resource Management, Pennsylvania State University, University Park, Pennsylvania, May 1996.

Maiolo, John and Paul Tschetter. *Social and Economic Impacts of Coastal Zone Development on the Hard Clam and Oyster Fisheries in North Carolina.* A report to the University of North Carolina Sea Grant Program. 1983.

Maril, Robert Lee. *Texas Shrimpers: Community, Capitalism, and the Sea.* College Station: Texas A&M University Press, 1963.

———. *Cannibals and Condos: Texans and Texas Along the Gulf Coast.* College Station: Texas A&M University Press, 1986.

McCoy, E.G. *Migration Growth Mortality of the North Carolina Pink and Brown* Penaid *Shrimps.* Special Scientific Report Number 15. Raleigh: Division of Commercial and Sportfisheries, North Carolina Department of Conservation and Development, 1968.

———. *Dynamics of North Carolina Shrimp Populations.* Scientific Report Number 21. Raleigh: Division of Commercial and Sportfisheries, North Carolina Dept. of Conservation and Development, 1972.

McKenzie, M.J. "Description of Industry: Harvesting Sector." In *The Shrimp Fishery of the South Atlantic United States: A Management Planning Profile,* Technical Report Number 5, edited by Dale Calder, et al.. Charleston: South Carolina Marine Resources Center, 1974.

North Carolina Division of Marine Fisheries. *Shrimp and Crab Trawling in the North Carolina's Estuarine Waters.* Morehead City, North Carolina: North Carolina Division of Marine Fisheries, 1999.

Orbach, M.K., and Jeffrey C. Johnson. "The Transformation of Fishing Communities: A Public Policy Perspective." In *Marine Resource Utilization: A Conference on Social Science Issues.* Proceedings of a conference presented by the Mississippi-Alabama Sea Grant Consortium, Mobile, Alabama, May 1988.

Pate, P.P. *Estimation of North Carolina Shrimp Harvest and Gear Utilization by Recreational and Commercial Fishermen.* North Carolina Division of Marine Fisheries. Project 2-275-R. Morehead City, North Carolina: North Carolina Division of Marine Fisheries, n.d.

Pollnac, R.B. "Sociocultural Aspects of Technological and Institutional Change Among Small-Scale Fishermen." In *Modernization and Marine Fisheries Policy,* edited by J.R. Maiolo and M.K. Orbach. Ann Arbor: Ann Arbor Science (Butterworth Group), 1982.

Prioli, Carmine, and Edwin Martin. *Hope for a Good Season.* Asheboro, North Carolina: Down Home Press, 1998.

Purvis, C., and E.G. McCoy. *Population Dynamics of Brown Shrimp in Pamlico Sound.* Special Scientific Report No. 24. Raleigh: Division of Commercial and Sportfisheries, North Carolina Department of Conservation and Development, 1975.

Sabella, James, Marcus Hepburn, and Rick Dixon. "Aspects of Family and Kinship in a North Carolina Community: A Comparative Study." *Maritime Politics and Management* 6 (1979): 2.

Smith, M.E. "Fisheries Management: Intended Results and Unintended Consequences." In *Modernization and Marine Fisheries Policy,* edited by J.R. Maiolo and M.K. Orbach. Ann Arbor: Ann Arbor Science (Butterworth Group), 1982.

South Atlantic Fishery Management Council (SAFMC). *Draft Amendment 2 to the Fishery Management Plan for the Shrimp Fishery of the South Atlantic Region.* Charleston: South Atlantic Fishery Management Council, 1995

———. *Fishery Management Plan for the Shrimp Fishery of the South Atlantic Region.* Charleston: South Atlantic Fishery Management Council, 1993.

——— *Profile of the* Penaid *Shrimp Fishery in the South Atlantic.* Charleston: South Atlantic Fishery Management Council, 1981.

South Carolina Division of Marine Resources and Wildlife. *Profile of the* Penaeid *Shrimp Fishery in the South Atlantic.* Charleston: South Atlantic Fishery Management Council, 1981.

Steinbeck, S., B. Gentner, and J. Castle. "Economic Impacts of Marine Recreational Angling in the United States: Selected Results from North Carolina." Paper presented at the spring meeting of the Southern Division, American Fisheries Society, Wilmington, North Carolina, February 14–16, 2003. NOAA Technical Report Series, Summer 3003.

Stick, David. *Graveyard of the Atlantic. Shipwrecks of the North Carolina Coast.* Chapel Hill: University of North Carolina Press, 1952.
———. *The Outer Banks of North Carolina.* Chapel Hill: University of North Carolina Press, 1958.
Still, William. "A History of the Shrimping Industry in North Carolina." *American Neptune* (Fall 1987):257–74.
Street, Michael, and Alice Williams. "Hurricane Floyd Commercial Fishermen's Relief Grant Program." Unpublished draft final report for the North Carolina Division of Marine Fisheries, May 2001.
U.S. Department of Commerce, National Oceanic and Atmospheric Administration, National Marine Fisheries Service. *Fisheries Statistics of the United States, 2001.* Washington, D.C.: U.S. Government Printing Office, 2002.
Whitehead, John, Bob Edwards, Marieke Van Willligen, John Maiolo, Kenneth Wilson, and Kevin Smith. "Heading for Higher Ground: Factors Affecting Real and Hypothetical Hurricane Evacuation Behavior." *Environment Hazards* 2 (2000): 133–142.
Wilson, C. *North Carolina Recreational Use of Commercial Gear Pilot Survey. Progress Report for the 2001 Survey.* Washington, North Carolina: North Carolina Division of Marine Fisheries, 2003.
Young, Claiborn. *Cruising Guide to Coastal North Carolina.* Elon, North Carolina: Watermark Publishing, 2003

About the Author

After receiving his Ph.D. in sociology from Penn State in 1965, John Maiolo taught and conducted social-policy research at Notre Dame and Indiana Universities. In 1975, he became chair of the combined Sociology and Anthropology Department at East Carolina University (ECU) and later supervised the development of ECU's Department of Economics. He was also instrumental in developing ECU's teaching, research, and service programs that focus on coastal-zone issues and fisheries management. He was awarded the title of professor emeritus upon his retirement from that institution in 2000. He has published several books, including one on the impacts of Hurricane Floyd, and numerous articles on this and other topics. He has also served as a consultant for private companies and governmental regulatory agencies, including the prestigious National Academy of Sciences, the South Atlantic Fishery Management Council, the Atlantic States Marine Fisheries Commission, and the North Carolina Division of Marine Fisheries. His research on the shrimp fishery was first undertaken in 1978 and has continued since then.